# Cold Hearths and Barren Slopes

Bina Agarwal

*Fuelwood: the primary source of inanimate energy in the Third World*

# Cold Hearths and Barren Slopes

## The Woodfuel Crisis in the Third World

Bina Agarwal

**Zed Books Ltd.**
**57 Caledonian Road**
**London N1 9B4**

*Cold Hearths and Barren Slopes* was first published in the United Kingdom by Zed Books Ltd., 57 Caledonian Road, London N1 9BU, in 1986.

Copyright © Institute of Economic Growth, Delhi, India, 1986

Printed in India by Allied Publishers Private Ltd., New Delhi

**British Library Cataloguing in Publication Data**

Agarwal, Bina
Cold hearths and barren slopes : the woodfuel crisis
in the Third World

1. Fuel  2. Power resources  3. Forests and forestry —
Economic aspects
 I. Title
 333.75'13          HD9750.6

ISBN 0—86232—539—0
ISBN 0—86232—540—4 Pbk

To all those striving to restore
the balance between people and nature

# FOREWORD

Many Third World countries are today faced with an energy crisis which relates not only to non-renewable fossil fuels, but also to the rapid depletion of a renewable resource—wood—which provides the principal, and often the only source of fuel to the majority of people in these countries. However, with growing deforestation and the consequent diminishing supply of wood-based fuels (primarily fuelwood and charcoal), large sections of the populations, especially of Asia and Africa, are facing a crisis.

Dr. Bina Agarwal, in this book, provides a comprehensive and detailed analysis of the causes and implications of the woodfuel energy crisis, as well as the efforts made towards, and difficulties encountered in promoting woodfuel innovations—viz., improved wood-burning stoves, forestry schemes, and improved charcoal kilns—as solutions to the crisis. The study notes that the major question of how to ensure that these innovations reach and benefit the rural poor, who are the most severely affected, has remained largely unexplored, and attempts to provide some answers. It highlights how the socio-economic context within which rural development is being undertaken in the Third World has implications both for the nature of this crisis and for the success of attempts to deal with it. Past experiences of development programmes directed at the rural poor are also brought to bear on the subject. These features (among others) make this study different from many existing ones which have tended to examine the woodfuel issue and related programmes somewhat in isolation from the more general experience of attempts to tackle poverty and promote rural development.

The study emphasizes the absence, in the short run, of viable alternatives to wood as the main source of domestic fuel in most parts of the rural Third World. It notes that the rural poor have to depend almost entirely on forests and common land for their fuelwood supply, and highlights the nature of privations suffered by them and especially by the women (who are usually the primary gatherers of fuelwood) because of shortages. At the same time, the study questions the popular assumption that the foraging of fuelwood by the rural poor is the main cause of the shortages, and presents several alternative explanations.

A major part of the book is devoted to identifying the socio-economic factors which affect the successful implementation of woodfuel programmes (especially those relating to improved wood-burning stoves and tree planting) for increasing the wood-energy available to the poorest sections. For this purpose the study also draws upon the experience of promoting other rural innovations, and seeks to conceptually incorporate these experiences by drawing up a typology of rural innovations in terms of their technical, economic and social characteristics. This typology, which also provides an analytical framework for understanding the diffusion process, is seen to indicate that innovations that require adaptation to user needs, which entail a financial cost but provide little direct financial benefit, and which are aimed at those who are economically and socially disadvantaged, are unlikely to be accepted through a market-oriented approach to promotion. Woodfuel innovations are noted to be characterized by one or more of these features and therefore likely to need a participatory approach to diffusion.

Moving from *a priori* reasoning to actual experiences of programmes for promoting improved wood-burning stoves, forestry schemes and improved charcoal kilns, especially in Asia and Africa, Dr. Agarwal notes that few such attempts have succeeded. She offers a strong critique, illustrated by numerous examples, of the shortcomings and inappropriateness of most such attempts. One of the significant causes of scheme failure is identified as the 'top-down' approach to innovation diffusion and scheme implementation. For instance, it is noted that technologies such as improved wood-burning stoves are often designed outside the social and cultural context in which they are to be used, and then introduced with little adaptation. Hence the specificity of user needs is not taken into account, leading to a high rate of stove rejection or a discontinuation of use after initial adoption. Again in planning social forestry schemes, inadequate consideration is given to the community's land use priorities, the existing land tenure and ownership patterns, and the tree species most suited to the varied needs of the people.

Dr. Agarwal makes a strong and persuasive case for user-involvement in the conceptualization, design and implementation of woodfuel schemes. Special stress is laid on the need to involve the rural poor and particularly the women. Examples of recent success stories, where alternative 'participative' methods for the promotion of rural innovations have been used, add considerable strength to the arguments presented in the book. But it is noted that where socio-economic inequalities (of class, caste, gender etc.) are high, these are likely to act as barriers to bringing about such an involvement.

The question of the appropriate methodology for evaluating woodfuel programmes is also discussed, and the importance of a participatory approach in data gathering and evaluation emphasized. It is further highlighted that much-used tools of project evaluation, such as the social cost-benefit analysis, involve value judgements which need to be made more explicit, and a special weightage given to distributional considerations in such exercises, especially when planning social forestry schemes.

This book, with its analytical insights, multidisciplinary approach and extensive geographic coverage, should be of considerable interest and use to a wide cross-section of people: academic researchers, teachers and students of social sciences and environmental sciences, policy makers and administrators, as well as grass-roots activists who are dealing directly with problems of poverty and fuel shortages.

Dr. Bina Agarwal's analysis, in contrast to that of many others, places the woodfuel issue in the wider developmental context. Her study is an important contribution to the literature, not only on renewable energy and forestry, but also on development issues in general.

**K. Krishnamurty**
*Director*

Institute of Economic Growth
Delhi, 1985

# PREFACE

This book is a substantially revised and updated version of a research monograph written in 1980, when I spent a year as a Research Fellow at the Science Policy Research Unit (SPRU), University of Sussex, UK. At that time the woodfuel issue was still relatively unresearched, and I had to depend a good deal on emerging unpublished material. In terms of policy too, in many Third World countries, a recognition of and concern for the problem was just beginning to be expressed. Since then, evidence of a deepening woodfuel crisis has grown, as has the spread of schemes that seek to alleviate it. Over a hundred programmes for promoting improved wood-burning stoves have been launched all over the Third World; and numerous schemes for encouraging tree planting, especially under the banner of social forestry, have also been initiated. Large amounts of international aid and national funds are flowing into such efforts.

Yet the approach to most such programmes continues to be a familiar one—a heavy reliance on a largely technological solution to a problem that is in fact rooted in the socio-politics of poverty and structural inequalities, and the associated maldistribution of scarce natural resources between uses and users. The method of programme implementation also typically continues to be 'top-down', usually with little involvement and participation of the rural poor, and with a high dependence on existing bureaucratic set-ups, such as the government rural extension networks and forestry departments, for promoting these schemes. Perhaps not surprisingly, the acceptance of improved wood-burning stoves in most programmes continues to be very low, and village woodlot schemes requiring community cooperation typically fail to take off. The critique of woodfuel programmes that I had spelt out in my SPRU monograph appears even more valid today—strengthened and substantiated by many more instances of usually predictable scheme failures.

An additional point of concern is that in several countries, State policies are leading on the one hand to the alienation and marginalization of tribal and poor rural communities from access to forest produce, and on the other hand to the spread of tree planting for commercial use (often in the name of social forestry), even at the cost of destroying natural forests and diverting fertile land away from crop cultivation. Implicit in such policies is also a separation of production considerations from those of distribution. However, on the positive side, there is a glimmer of hope in the emergence in some countries of ecology-related protest movements among tribal and other rural communities, and of urban groups seeking to raise environmental consciousness in different forums. In many such efforts, people are seeking not merely to defend themselves against the negative onslaught of State policies but also to act as agents of positive change for restoring the lost balance between society and nature. In one such movement—the *Chipko Andolan* in the hills of Uttar Pradesh in India—the songs of a folk poet have become the cry of an emerging consciousness:

> *'Come, arise, my brothers and sisters*
> *Save this mountain . . .'*
> *'Come plant new trees, new forests*
> *Decorate the earth.'*

A description of this movement, and of several other success stories, where a participative approach to scheme implementation was followed, are among the many additions and changes that have gone into the updating and rewriting of this book in its present form. They temper the note of pessimism that prevailed in the earlier monograph because of the predominance of examples of failure. Also new material and issues have been incorporated, including the issue of social forestry which has today become part of a significant debate, especially in India.

Being based in India these past few years, my access to recent material from other Third World countries has been limited somewhat. Hence many (but by no means all) of the illustrative examples of the most recent period, relate to India and the subcontinent. However, these examples support the analysis and understanding that emerges from the earlier literature relating to other parts of the globe, and can be seen to have a relevance beyond their country-specificity.

The writing of both the SPRU monograph and the present version would not have been possible without the support of many friends and associates. At SPRU, Andrew Barnett and Charles Cooper followed the study at every stage with meticulous comments and an unflagging interest. This was invaluable to me when working on the monograph, and I am deeply indebted to both of them. I also greatly appreciate the detailed and incisive comments given by Martin Bell and Kurt Hoffman, and the many helpful suggestions made by Mike Howes and Ariane Van Buren on an earlier draft. To Robert Chambers I owe a special thanks for providing me with several interesting and relevant papers which proved most useful in updating the study. The SPRU monograph was sponsored by the United Kingdom Tropical Products Institute and I am grateful to them for funding it.

The rewriting and updating of the book in its present form took several months and was undertaken in Delhi after I joined the Institute of Economic Growth (IEG). I am grateful to Dr. K. Krishnamurty, Director of IEG, for his institutional support. The help provided by the staff of the libraries of the IEG, the Food and Agricultural Organisation of the United Nations (Delhi), and the National Council of Applied Economic Research (Delhi), is also gratefully acknowledged. In particular, I would like to thank Anil Agarwal of the Center for Science and Environment for giving me generous access to the material available at the Center, and to Sunita Narain and Priya Deshingkar for their help in locating that material. I am also grateful to Ravi Vyas of Macmillan (India) and Ritu Menon of Kali for Women (India) for their editorial suggestions on the draft manuscript. I am indebted to Shri Sunderlal Bahuguna for sharing with me, during our brief meeting, two of the most recent songs of the *Chipko* Movement and helping me in my translation of them into English. Valuable help was also provided by Sudhir Dhaundiyal in translating some of the older *Chipko* songs from Garhwali to Hindi (subsequently translated by me into English). I am indebted too to several organisations and friends for the illustrative photographs, and especially to Smitu Kothari for his keen interest and help in the pictorial layout and cover design.

Most of all I would like to thank my father, Shri S.M. Agarwal, who read the last two drafts of the book with meticulous care and offered many useful comments. In particular, his capacity to immediately spot repetitions and inadequately supported or unclearly expressed arguments, and his suggestions on the overall organisation of the book, have contributed greatly to its clarity and readability.

Ultimately, the production of any piece of research depends a good deal on the quality of human interaction and warmth from those one is in everyday contact. It is impossible for me to individually name all those who have so contributed to my work: it includes my friends and colleagues at the University of Sussex who made my stay there an enriching experience, intellectually and emotionally; my family which has always been highly supportive; and my friends in Delhi who extended their patience and understanding when I hibernated over many evenings and weekends to complete the book.

Bina Agarwal
Delhi, 1985

# CONTENTS

# CHAPTER I

# THE NATURE OF THE WOODFUEL PROBLEM

*Remember those forests of oak and rhododendron,*
*fir and spruce,*
*those trees of pine and deodar*
*that have vanished?...*

*The trees near the streams have been felled,*
*the rivers have run dry;*
*the wild fruit, the herbs are gone,*
*the berries, the wild vegetables have disappeared ....*

*With the felling of trees landslides have started;*
*our fields, barns and homes are all washed away.*
*Where once there were lush forests*
*there is now sparseness. ...*

—from three Garhwali folk songs
by Ghan Shyam Shailani[1]

The overwhelming preoccupation with and publicity given to the world oil crisis over the past several years, contrasts sharply with the relatively little recognition given to another energy crisis—a quiet one—which pre-dates that of oil. This is the crisis of woodfuel* shortages facing a vast proportion of people in the Third World, for whom it is the principal, and often, the sole source of inanimate energy.**

---

*The term 'woodfuel' refers throughout to wood-based fuel (essentially firewood and charcoal).
**This includes all energy sources other than human and animal.

The shortages are manifest in the longer hours spent, especially by women and children, for gathering fuel and fodder, in families going hungry because they do not have enough firewood to cook the food, in the rising prices of firewood in cities, in the shift to cattle dung and other substitute fuels for domestic use, in the growing confrontations between forest communities and forestry officials, and so on.

As forests in the Third World are beginning to disappear (attributable in large measure to historical and ongoing malpractices, and State policies) more and more people are being forced to depend on fewer and fewer tree resources for fuel. Anecdotal evidence on people's growing desperation is often dramatic. One observer notes how with the gradual denudation of the Sahelian countryside people have started cutting down saplings (Floor, 1977). Another mentions seeing people strip the bark off the trees that line the roads in Peshawar, Pakistan (Eckholm, 1975: 6). A third makes the following comment about the hills of Nepal: 'Nowhere can there be seen a tree or bush unscarred by axes, knives and browsing domestic animals. The imprint of people searching for fuel and fodder is to be seen everywhere' (Hughart, 1979: 28). Outside Ougadougou, in

*Mark Edwards/Earthscan*

*Around the village—an expanding wasteland: Kordofan, Western Sudan*

Niger, the land is noted to have been stripped of trees for 45 miles in all directions (French, 1978: 1), and the same is observed to be true for Niamey (capital of Niger) around which there is now a virtual desert in a 70 kilometer radius (Spears, 1978: 3).

In the Himalayan foothills of Nepal a journey to gather firewood and fodder took an hour or two a generation ago—today it is observed to take a whole day (Eckholm, 1975: 7). In Bara (Sudan) where ten years ago fuelwood was said to have been available after a 15 to 30 minute walk from the village, women now have to walk for at least one to two hours (Digerness, 1977: 15). In some villages of Gujarat, in India, where the surrounding forests have been completely denuded, women spend long hours collecting weeds and shrubs and digging out the roots of trees (Nagbrahman and Sambrani, 1983: 36, 37). Inadequacies of fuel are driving people in several regions to shift to foods that are less fuel-consuming but of lower nutritional value, or to miss some meals altogether and go hungry. As one author puts it: 'None of the principal foodcrops of the tropics is palatable unless it has been cooked first. Lack of fuel can be as much a cause of malnutrition as the lack of food' (Poulsen, 1978: 13).

Yet the 'discovery' of a woodfuel crisis by policy makers in the Third World countries is not merely recent but also incidental—incidental on the one hand to a general assessment by these countries of their energy 'options' and on the other to the connection made, often unjustifiably, between deforestation (with its unignorable, adverse ecological consequences) and the gathering of wood for fuel.

The discovery, in turn, has brought in its wake the offering of a spate of 'solutions' in the form of improved wood-burning devices (especially domestic cooking stoves), a variety of tree-planting schemes, and improved wood-conversion hardware (especially charcoal kilns). These three have broadly been termed here as 'woodfuel innovations'.* Yet most of the attempts to promote these innovations have so far been ineffective.

The major part of the present study is devoted to an examination of the problems faced in the diffusion of these innovations and to an identification of the factors underlying the failure of many such promotional efforts. It is shown here that the nature of the woodfuel

---

*The term 'innovation', here and subsequently, has been used in a broad sense to include both objects and practices perceived as being new by an individual or group, even if it has previously been in existence or in use elsewhere. Hence, improved stoves and kilns and tree-planting schemes could all be termed innovatious in this sense.

crisis, the causes underlying it, as well as the possibilities of alleviating it are linked to a complex set of social, economic and political factors. And even if the attempted solutions are confined to schemes for promoting hardware and reforestation, a range of problems, stemming from these factors, are likely to be encountered in the implementation of such schemes. These are problems which cannot be solved in the laboratory, and which highlight the need for a radical departure from the typical 'top-down' approach to innovation diffusion, and for some basic institutional and structural changes.

The case of woodfuel innovations is considered here within the general context of the diffusion of innovations in rural areas, and is seen as being in many ways similar to, and in many others distinct from that of other rural innovations. Given the overall dearth of case study literature relating specifically to the diffusion of woodfuel innovations, an examination of the experience of other innovations is not only useful, but often crucial to an understanding of the problems relating to woodfuel innovations.

However, the search for an effective solution to the woodfuel crisis, it is argued here, cannot be confined only to the issue of using available wood supplies more efficiently (hardware), or of increasing the supplies of wood available (forestry). It must also take account of the need to alter the existing distribution between uses and users, both of wood and of non-wood-related energy sources.

The book is divided into seven chapters and an appendix. The rest of the present chapter attempts to 'define' the nature of the woodfuel problem and to question some currently popular explanations for woodfuel shortages. It also spells out the issues relating to a search for 'solutions'. Chapter II summarizes the technical aspects of the main woodfuel-related options—that is, improved wood-burning stoves and tree-planting schemes—and also briefly describes improved charcoal-producing hardware. Chapters III to V which constitute the body of the book examine the socio-economic factors affecting the adoption of these options. In Chapter III, the literature relating to the diffusion of rural innovations in general is reviewed, and the main approaches put forward by various scholars identified. An attempt is also made to categorize different types of rural innovations by their social, economic and technical characteristics, and to apply these categories to woodfuel innovations, for establishing, on *a priori* grounds, which factors are most likely to affect the diffusion of such innovations. In Chapters IV, V and the Appendix, case-study literature relating to the actual experience with diffusing, respectively, improved wood-burning stoves, tree-planting schemes and improved charcoal kilns is considered. The noticeably larger number of studies which deal with programme failures, rather than successes, reflects the nature of the available literature (no selectivity has been applied here!), and is taken to be representative of the actual situation. Among other things, this literature is seen to establish a strong case for user-participation in the innovation and scheme-implementation process. At the same time, the likely constraints to such participatory involvement, in the form of inequalities in the distribution of economic and political power between classes and social groups, are also discussed. Chapter VI focuses on the methodology for evaluating diffusion schemes. Chapter VII, in conclusion, draws together the major issues emerging from the study and spells out their implications.

Let us here begin by considering what the nature of the woodfuel problem is and in what way, for whom, and why wood shortages have acquired the dimensions of a crisis.

## 1. The Importance of Wood as a Source of Fuel

The 1973 oil price hike initiated by the Organisation of Petroleum Exporting Countries (OPEC) provoked a world-wide alarm about an impending 'energy' crisis. The message was treated as ominous, especially by the oil-importing Third World countries, since it meant that the foreign exchange costs of such imports would become increasingly prohibitive.[2] Added to this was the fear that even at the going price adequate supplies would not long be available. The same was seen to hold true for other fossil fuels.*

The realisation that there are physical limitations to the world availability of such fuels (quite apart from the financial costs) has led to a search for alternatives among renewable energy sources,** not merely to substitute for existing requirements but also to provide for the future. And one of the most important sources of renewable energy presently in use in large parts of the Third World is woodfuel. This includes wood burnt directly as a fuel (here termed 'firewood' or

*Coal, oil and natural gas.
**These cover solar, hydro-electric, wind, ocean-thermal, tidal, biomass (including wood) and nuclear.

'fuelwood') as well as fuels derived from wood, such as charcoal, producer gas, water gas, methanol and gasoline. Among these derived fuels, charcoal is the most important in the present context.

Unfortunately, no precise macro-estimates exist of the use of woodfuel in different countries.[3] Most studies rely on the figures published by the United Nation's Food and Agricultural Organisation in its *Annual Yearbooks of Forestry Products*. These figures relate to the annual production of fuelwood plus charcoal (for domestic and other uses) which is typically assumed to equal annual woodfuel consumption. For some countries the figures given are either based on FAO's own estimates or obtained from unofficial sources. For others, they are official figures supplied by the countries and would usually cover only recorded removals from forests. Since the fuelwood consumed domestically is often gathered in the form of twigs and branches from forest floors or from trees located on open roads and in fields[4]—removals which go largely unrecorded—the official figures of production would tend to underestimate actual consumption. This underestimation may be quite substantial in some regions, as was noted by Openshaw (1971) for Tanzania, when he compared the figures for recorded production of fuelwood and charcoal in 1970 with the consumption figures arrived at on the basis of survey data. A similar observation is made by Earl (1975: 58) for the 1970 FAO figures for Nepal.[5] Further, although FAO gives country-specific figures for imports and exports of fuelwood plus charcoal, the world aggregates do not balance: the 1982 world export figure was only 74 per cent of the world import figure (FAO, 1984). Hence the FAO woodfuel-related statistics need to be treated with caution.[6] Nevertheless, these figures are still generally considered the best available at the macro-level, and in the absence of better information they continue to be useful in providing broad pointers and giving an overall picture.

On the basis of the FAO figures, it is estimated that currently two-thirds of all inanimate energy in Africa, one-third in Asia, and one-fifth in Latin America comes from fuelwood and charcoal (Arnold and Jongma, 1977: 2). Also, Table 1, which gives data for 1982, indicates that in many countries such as Benin, Chad, Ethiopia, Nepal, Rwanda, Uganda and Upper Volta, firewood and charcoal together provide close to 90 per cent or more of the total inanimate energy used, and for the majority of other countries in Africa, and many in Asia and Central America, the figure is well over 50 per cent.[7]

*A forest in Latin America (above); the same region five years later (below)*

**Table 1: Commercial and woodfuel energy consumption in the Third World (1982)**

| Countries | Commercial energy per capita kg. CE* | Woodfuel† consumption per capita | | Woodfuel as a per cent of total energy (commercial+woodfuel) |
|---|---|---|---|---|
| | | Cu.m | kg. CE | |
| *Africa* | | | | |
| Angola | 126.0 | 0.96 | 319.0 | 71.7 |
| Benin | 39.7 | 1.05 | 348.4 | 89.8 |
| Central Africa | | | | |
| Republic | 41.7 | 1.06 | 353.0 | 89.4 |
| Chad | 21.5 | 1.62 | 541.3 | 96.2 |
| Ethiopia | 31.0 | 0.83 | 276.2 | 89.9 |
| Ghana | 122.3 | 0.60 | 198.8 | 61.9 |
| Ivory Coast | 287.0 | 0.78 | 260.4 | 47.6 |
| Kenya | 102.6 | 1.48 | 492.3 | 82.8 |
| Madagascar | 66.8 | 0.59 | 197.4 | 74.7 |
| Malawi | 50.0 | 0.90 | 301.3 | 85.8 |
| Mali | 29.6 | 0.58 | 194.2 | 86.8 |
| Niger | 47.3 | 0.58 | 191.7 | 80.2 |
| Nigeria | 185.6 | 0.84 | 278.7 | 60.0 |
| Rwanda | 21.3 | 1.11 | 370.9 | 94.6 |
| Senegal | 199.7 | 0.55 | 183.1 | 47.8 |
| Sudan | 84.0 | 1.75 | 583.7 | 87.4 |
| Uganda | 25.3 | 1.77 | 590.2 | 95.9 |
| Upper Volta | 31.4 | 1.04 | 346.7 | 91.7 |
| Zaire | 68.9 | 0.91 | 303.5 | 81.5 |
| Zambia | 389.2 | 0.87 | 289.6 | 42.7 |
| Zimbabwe | 636.1 | 0.96 | 321.4 | 33.6 |
| *Asia/Pacific* | | | | |
| Afghanistan | 50.1 | 0.30 | 98.9 | 66.4 |
| Bangladesh | 49.8 | 0.32 | 108.5 | 68.5 |
| China | 581.3 | 0.16 | 51.7 | 8.2 |
| India | 200.6 | 0.29 | 96.8 | 32.5 |
| Indonesia | 234.9 | 0.74 | 245.1 | 51.1 |
| Korea | | | | |
| (Republic of) | 1439.0 | 0.18 | 58.8 | 3.9 |
| Malaysia | 985.8 | 0.50 | 165.2 | 14.4 |
| Nepal | 11.2 | 0.90 | 298.6 | 96.4 |
| Pakistan | 225.8 | 0.21 | 68.7 | 23.3 |
| Philippines | 329.6 | 0.55 | 182.4 | 35.6 |
| Sri Lanka | 121.9 | 0.49 | 163.8 | 57.3 |
| Thailand | 364.4 | 0.73 | 243.5 | 40.0 |
| Papua (New | | | | |
| Guinea) | 293.2 | 1.78 | 594.4 | 67.0 |

*South and Central America*

| | | | | |
|---|---|---|---|---|
| Brazil | 704.8 | 1.25 | 417.5 | 37.2 |
| Chile | 908.9 | 0.50 | 167.9 | 15.6 |
| Colombia | 897.2 | 0.50 | 168.2 | 15.8 |
| Cuba | 1440.4 | 0.29 | 95.4 | 6.2 |
| Guatemala | 217.5 | 0.84 | 279.3 | 56.2 |
| Honduras | 233.2 | 1.03 | 342.1 | 59.5 |

*Notes*: (1) * CE = coal equivalent.

   † Woodfuel consumption figures relate to fuelwood+charcoal.

(2) The total energy estimates do not include other traditional sources of energy such as cattle dung, crop residues, etc. because of the absence of micro-estimates for these sources.

(3) Per capita consumption has been computed by dividing:

   (a) available figures of total commercial energy in kg. CE, and total woodfuel energy consumption in cubic metres; and

   (b) computed figures of total woodfuel energy consumption in kg. CE

   by available estimates of populations in mid-1982.

(4) Conversion factor used for woodfuel: 1 cubic metre = 0.33 metric tons CE. (This is taken from *Energy Statistics Yearbook 1982*; p. xxvi.)

*Sources*: (1) Data on total commercial energy taken from United Nations (1984): *Energy Statistics Yearbook 1982*, Dept. of International Economic and Social Affairs, Statistical Office, New York.

(2) Data on total woodfuel consumption taken from FAO (1984): *1982 Yearbook on Forestry Products*, FAO Forestry Statistics Series, Rome. (Note: Production is assumed as equal to consumption in this data source.)

(3) Data on population taken from World Bank (1984): *World Development Report, 1984*, Washington D.C.

Again, while there are no accurate macro-estimates of the proportion of woodfuel burnt directly as wood (i.e. firewood) and that burnt as charcoal etc., broad assessments suggest that firewood is by far the more important. From FAO estimates, the world production of charcoal in 1982 comes to only 6.6 per cent of the world woodfuel production in that year. A range of macro-studies indicate that the use of charcoal is predominantly concentrated in the urban areas. In rural areas, wood is usually burnt directly as a fuel in most households (see Fleuret and Fleuret, 1978; Howe, 1977; and Uhart, 1976).

The consumption of woodfuel in per capita terms varies greatly between the Third World countries as seen from Table 1. Leaving aside countries such as China, S. Korea and Cuba where woodfuel energy constitutes less than 10 per cent of the total energy consumed,

and basing our comparison on the others, we note from the table that
the lower end of the range is occupied largely by the countries of
South Asia, with per capita consumption in most being less than
0.5 m³, and being especially low in Bangladesh, India and Pakistan. The
upper end of the range is occupied largely by the countries in Africa,
many of which have a per capita consumption above 1.0 m³, and
some such as Chad, Sudan and Uganda, consume over 1.5 m³. In
between fall the majority of countries in Africa, S.E. Asia and South
and Central America, with consumption ranging between 0.5 and

### Table 2: Fuelwood* use by ecological regions (1980)
(in m³ per capita per year)

| | Fuelwood* | |
|---|---|---|
| | Needs | Availability |
| *Africa* (South of Sahara) | | |
| Arid and sub-arid areas | 0.5 | 0.05 to 0.01 (AS) |
| Mountainous areas | 1.4 to 1.9 | 0.5 to 0.7 (AS) |
| Savanna areas        (a) | 1.0 to 1.5 | 0.8 to 0.9 (D) |
|                     (b) | | 1.8 to 2.1 (PD) |
| High forest areas   (a) | 1.2 to 1.7 | 1.8 to 2.0 (PD) |
|                     (b) | | 5.0 to 10.0 (S) |
| *Asia* (Far East) | | |
| Mountainous areas | 1.3 to 1.8 | 0.2 to 0.3 (AS) |
| Indo-gangetic plains (S. Asia) | 0.2 to 0.7 | 0.15 to 0.25 (D) |
| Low land areas in S.E. Asia | | |
|   (Plains and islands) | 0.3 to 0.9 | 0.2 to 0.3 (D) |
| High forest areas | 0.9 to 1.3 | 1.0 to 6.0 (S) |
| *Latin America* | | |
| Andean plateau | 0.95 to 1.6 | 0.2 to 0.4 (AS) |
| Arid areas | 0.6 to 0.9 | 0.1 to 0.3 (AS) |
| Semi-arid areas | 0.7 to 1.2 | 0.6 to 1.0 (D) |
| Sub-tropical and | | |
|   temperate areas | 0.5 to 1.2 | 1.9 to 2.3 (PD) |
| Abundant forest areas | 0.5 to 1.2 | 2.5 to 10.0 (S) |

*Note*: *Wood for charcoal is included here. Hence, strictly in terms of the definition
used in this book, these figures would relate to 'woodfuel', not just to
'fuelwood'.
AS: Acute Scarcity; D: Deficit; PD: Prospective Deficit; S: Satisfactory.
*Source*: FAO (1981): *Map of the Fuelwood Situation in Developing Countries—
Explanatory Note*, FAO, Rome.

1.0 m³. One of the important factors influencing these variations in consumption among countries where woodfuel is an important energy source, is the relative availability of wood for fuel in different ecological regions. It can be seen from Table 2 that the per capita availability of fuelwood varies from levels several times above estimated needs in the high forest areas of the world, to much below needs in the arid and sub-arid zones.

Locational availability also leads to significant variations in consumption *within* countries. In Nepal, for instance, it has been noted that people who migrated to the forested *Terai* plains where wood is relatively plentiful, consume twice as much firewood as those left behind in the forest-denuded hills (Earl, 1975: 113). Again, in Tanzania, households in villages near wooded areas are noted to consume over three times the woodfuel consumed by households in villages with little or no surrounding woodland (Arnold, 1978: 8). Likewise, in India it was found in a survey that while villages located inside or near a forest met 100 per cent of their fuel needs from the forest, this percentage decreased progressively as the distance of the villages from the forest increased, dropping to almost nil at a distance of 15 km (Arnold, 1978: 10). Within regions again, there are variations as households differ in their cooking styles, type of meals cooked, family size and especially income levels. From Table 3, which gives the average consumption of firewood per household by income class in different environmental contexts in north and south India, it is seen that consumption increases consistently as income levels rise, in each region.

The domestic consumption of fuelwood energy constitutes its primary use. In Pura village of Karnataka (India), for example, 90 per cent of the firewood consumed is noted to be for domestic purposes, 4 per cent for cottage industries and 6 per cent for other industries (Ravindranath, *et al.*, 1978). Again, country-level assessments for several countries indicate that 75-89 per cent of the fuelwood consumed is in the rural household sector, 7-9 per cent being used in cottage industries and 2 to 18 per cent in other industrial and service sectors.[8] Within the domestic sector, the primary use of fuelwood energy is for cooking. For instance, surveys conducted in Gambia and Thailand indicate that cooking (including water heating, baking and brewing beverages) accounts for over 75 per cent of total firewood consumption (Hughart, 1979). In Pura village (India) 82 per cent of total firewood consumed by the community is found to be for

**Table 3: Annual consumption of firewood per household by region and income class in rural India**
(kg CR*)

| Income class Rs./annum | Region | | | |
|---|---|---|---|---|
| | Hills | Plains | Deserts | All regions |
| *Northern India* (1975-76) | 1103 | 466 | — 822 | 551 |
| Upto 3000 | 926 | 293 | 624 | 374 |
| 3001 to 6000 | 1310 | 774 | 1072 | 862 |
| 6001 to 9000 | 1663 | 795 | 1571 | 905 |
| Over 9000 | 2320 | 1362 | 1661 | 1454 |
| | Hills | Plains | Coastal | All regions |
| *Southern India* (1979-80) | 549 | 584 | 362 | 489 |
| Upto 3000 | 534 | 568 | 294 | 455 |
| 3001 to 8000 | 571 | 604 | 484 | 549 |
| Over 8000 | 806 | 1084 | 412 | 686 |

*Note*: *CR = coal replacement.
*Sources*: (1) NCAER (1981: 105): *Report on Rural Energy Consumption in Northern India*, Environment Research Committee, National Council of Applied Economic Research, New Delhi.
(2) ITES (1981: 424): *Rural Energy Consumption in Southern India*, Institute of Techno-Economic Studies, Madras.

cooking (Ravindranath, *et al.*, 1978).

In most of the Third World, dependence on firewood as the primary source of fuel for domestic energy and particularly for cooking energy, is extremely high, especially in the rural areas. In rural Brazil, for example, firewood is noted to account for 79 per cent of all cooking fuel, while in rural Ghana and in parts of India (e.g. Pura village) it accounts for 100 per cent (DEVRES, 1980: 45, 46). Firewood may be supplemented by agricultural residues and animal wastes, but typically these are of relatively much less importance except where woodfuel shortages have left few other options.

According to one assessment of the total traditional energy (from firewood, crop residues and dung) consumed in Third World countries, 79 per cent is as firewood, 17 per cent as crop residues, and 4 per cent as dung; in some countries, as in Nigeria and Indonesia, the percentage for firewood is even higher—92 and above (Smil, 1979: 529). In general, for a large proportion of the people in the rural areas of the Third World (especially in Asia and Africa) firewood constitutes the main and, for some, the sole source of inanimate energy.[9] Given that in the majority of these countries 70 per cent or more of the populations are rural-based, this would account for a significant dependence by their people on firewood. All said, wood (whether burnt directly as firewood or after conversion to charcoal) may be seen to be the most important source of energy, especially domestic energy, in use in much of the Third World.[10]

Furthermore, a number of studies suggest that other renewable energy alternatives, such as solar and wind technologies or even biogas,[11] are not yet at a stage of development where they can be adapted successfully to adequately serve the needs that woodfuels currently serve in Third World countries. In India, for example, the Government's *Energy Policy Report* concludes that solar and wind power are not likely to provide viable options until at least the year 2000 (Prasad, 1979: 396). Also, there are noted to be a variety of specific problems, in addition to high costs, in adapting solar cookers for home use—such as the time of cooking, the adequacy of sunlight etc., which are seen to make them unsuitable in particular cultural and physical environments. An evaluatory study of efforts to promote solar cookers in rural India, undertaken by the National Physical Laboratory of India, is noted to have concluded: 'Since tradition in cooking methods plays a very major part, it is doubtful if much could be done in changing traditional methods . . . efforts made to introduce solar cookers in villages have completely failed' (quoted in Walton, *et al.*, 1978: 26).

Problems with the mass adoption of biogas have been found to be of a somewhat different nature.[12] At the level of the individual adopter they relate to the high cost of the plants, the inadequate number of cattle owned, or the total absence of cattle ownership in many households (3-4 head of cattle is considered to be a minimum need for a family-sized biogas plant), the absence of an assured water supply, and the requirements of technical knowledge, all of which usually make the plants viable only for the rich. At the community level, difficulties relate to the collection of biogas and the distribution of gas between users.

China is one of the few countries where family-based biogas plants have been adopted on a fairly extensive scale—it is estimated that in 1979 there were 7.2 million such plants in operation, and an average of 1 million new ones have been installed each year since 1975 (Van Buren, 1980: 7). The bulk of these plants are in the rural areas. Here a large number of households own a few pigs and there are no social taboos associated with the collection of nightsoil. Also, there are a range of plant designs which can be built with local materials by the people themselves. This means that very low costs are involved, unlike, say in India, where plants with pre-fabricated, galvanised iron domes have more commonly been experimented with, involving an expenditure of around Rs.3000 per plant for the user[13] (see Reddy and Prasad, 1977: 1484).

It could be argued that in countries such as India, the widespread adoption of biogas plants by the rural rich could still help the poor indirectly, by making more alternative fuels available for the latter. This argument, however, is weak on two grounds. First, it is far from inevitable that the use of biogas by the rich would increase the availability of other fuels for the poor, since the rich might (a) merely increase their overall energy consumption, with biogas adding to, rather than sustituting for, other sources; and (b) substitute biogas for commercial fuels such as kerosene, electricity, etc., rather than fuelwood. Second, if dung becomes valuable to the rich, the poor might no longer have the right to free dung collection; this would in fact increase their burden.

Hydro-power is another source of renewable energy—but here again its dependence in many regions on rainfall levels, together with the overall limitations in the widespread use of electricity for cooking in rural areas, are constraining factors.

Taken together, most studies looking at renewable energy options in the Third World, directly or indirectly indicate that, as at present, dependence on woodfuels (in current uses) is likely to continue for some time to come in most Third World countries.[14] This means that not only would existing household requirements have to be better satisfied (involving an increase over current low levels of consumption for many poor households), but that any rise in requirements due to say population increases would need to be provided for until such time as alternatives to woodfuels can be adopted widely.

## 2. Firewood Collection: From Where and By Whom?

In most rural areas, firewood has been and still largely tends to be non-monetized so that people usually have to depend on what they can themselves gather. By one estimate, for rural India, only 12.7 per cent of total firewood consumed is purchased, the rest being obtained either from one's own land or from the land of others (Government of India, 1982: 6). In Nepal, in some rural areas, (e.g. Pangua village) all firewood needs are met by self-collection, nothing is purchased (Bajracharya, 1983a). The same is true for some villages in Peru (Skar, 1982). In fact, the self-collection of firewood still appears to be the commonly observed pattern in most of the rural Third World (Arnold and Jongma, 1977).

Rural households with land can gather firewood from trees located on their own land, supplemented by crop residues, etc. The landless, however, have to depend on wood from common land, or where allowed to do so, gather it from other people's land by say contributing labour in return. As incomes decline, the dependency for fuel collection from sources other than one's own increases: Table 4 based on survey data illustrates this for south India..

Inequalities in land ownership patterns also lead to high inequalities in fuel availability between households. In Bangladesh,

## Table 4: Fuel collection from own sources by income class in rural south India (1979-80)

| Income class Rs./annum | Per cent of total households collecting fuel from own sources | | |
|---|---|---|---|
| | Firewood | Dung cake | Vegetable wastes |
| Less than 3000 | 33.8 | 69.1 | 43.8 |
| 3001-8000 | 70.2 | 92.6 | 77.9 |
| 8001 and above | 83.8 | 93.0 | 98.1 |
| All classes | 60.9 | 84.5 | 69.9 |

*Source*: ITES (1981: 176): *Rural Energy Consumption in Southern India*, Institute of Techno-Economic Studies, Madras.

Briscoe (1979: 628) noted, for example, that in the village he surveyed, 80 per cent of all fruit and firewood trees were owned by 16 per cent of the families, who also owned 55 per cent of the cropped land and 46 per cent of the cattle. Islam (1980: 73) again, on the basis of a survey of Bangladeshi villages, noted that 52.4 per cent of the trees were owned by 11.4 per cent of the households. A similar pattern is found in village studies in India: a six-village study in Karnataka highlights the close positive relationship between average landholding size and the quantity of firewood consumed in the household (ASTRA, 1981). In other words, access to land affects not only the household's access to food but also to the fuel used to cook it.

The collection of firewood in most parts of the Third World is done primarily (and sometimes exclusively) by women and children, with men usually (but not always) providing some supplementary labour. The actual time taken for collection varies in different regions according to the availability of tree resources, but in most cases it is a strenuous and time-consuming task. Table 5 brings together some of the existing studies which indicate the time taken and distances travelled in rural households for firewood collection. In a significant

F. Mattioli/ WEP (FAO)

*Ever lengthening treks: Mali*

number of cases the time is 3-4 hours per day or more. In some areas, as in the Sahel, women have to walk up to 10 km for this purpose; in Gambia it takes from midday to nightfall to gather an evening's supply, while in parts of India women spend five hours per day on an average, travelling 5 km or more over difficult mountainous terrain. Fleuret and Fleuret (1978: 318) note in the context of Tanzania: 'Every aspect of fires and fuels is the work of women in Kwemzitu, and no other task is considered to be as tiring or as demanding, or to have so little to show for itself'. In some parts of East Africa, the collection of firewood is considered appropriate only for married women, although in most countries, children, especially young girls, help their mothers (DEVRES, 1980: 21).

While the collection of firewood is primarily for self-consumption, yet increasingly as other sources of livelihood get eroded, selling firewood for an income is becoming common in some areas. The sellers are usually rural women belonging to the poorer households, who take the wood to nearby towns. In the Bara (Sudan), firewood sales represent the sole source of income in many cases (Digerness, 1977). This is also true in parts of India, as for example in certain areas of Bihar, where women of poor tribal households travel 8-10 km in search of firewood (which is usually procured illegally[15]), then catch a truck or train to the nearby town (Ranchi), spend the night at the station, and return with a meagre earning of Rs.5.50, on an average, for a headload of about 20 kg of wood (Bhaduri and Surin, 1980). 'Headloading' as it is called has become common here after a drought left many households destitute, some years ago. By one estimate 2-3 million rural people in India are headloaders and are spread across several States (Agarwal and Deshingkar, 1983). Wood is also sometimes converted into charcoal and sold by the rural poor for a livelihood in several Third World countries (DEVRES, 1980).

## 3. Implication of Shortages

The availability of wood, even though it constitutes a *renewable* resource, is limited, and becoming increasingly short in supply. Large tracts of land which were earlier thickly wooded, today lie barren. By one estimate there will be virtually no forests left in the Third World countries in 60 years time, in the absence of serious measures

**Table 5: Time taken and distance travelled for firewood collection by regions**

| Country | Region | Year of data | Firewood collection* | | Data source |
|---|---|---|---|---|---|
| | | | Time taken | Distance travelled | |
| *Asia* | | | | | |
| Nepal | Tinan (hills) | 1978 | 3 hr/day | n.a. | Stone (1982) |
| | Pangua (hills) | late 1970s | 4-5 hr/bundle | n.a. | Bajracharya (1983a) |
| | n.a. | n.a. | 0.62 hr/day | n.a. | Acharya and Bennett (1981) |
| India | Chamoli (hills) | | | | |
| | (a) Dwing | 1982 | 5 hr/day† | over 5 km | Swaminathan (1984) |
| | (b) Pakhi | | 4 hr/day | over 3 km | |
| | Gujarat (plains) | | | | |
| | (a) Forested | 1980 | Once every 4 days | n.a. | Nagbrahman and Sambrani (1983) |
| | (b) Depleted | | Once every 2 days | 4-5 km | |
| | (c) Severely depleted | | 4-5 hr/day | n.a. | |
| | Madhya Pradesh (plains) | 1980 | 1-2 times/week | 5 km | Chand and Bezboruah (1980) |
| | Kumaon Hills | 1982 | 3 days/week | 5-7 km | Folger and Dewan (1983) |
| | Karnataka (plains) | n.a. | 1 hr/day | 5.4 km/trip | Batliwala (1983) |
| | Garhwal (hills) | n.a. | 5 hr/day | 10 km | Agarwal (1983a) |
| Bangladesh | Chargopal | 1977 | 0.4 hr/day | n.a. | Cain. et. al. (1979) |
| Indonesia | Java | 1972-73 | 0.3 hr/day | n.a. | White (1976) |
| *Africa* | | | | | |
| Sahel | n.a. | c.1977 | 3 hr/day | 10 km | Floor (1977) |
| | n.a. | 1981 | 3-4 hr/day | n.a. | Ki-Zerbo (1981) |
| Niger | n.a. | c.1977 | 4 hr/day | n.a. | Ernst (1977) |

| Upper Volta | n.a. | n.a. | 4½ hr/day | n.a. | Ernst (1977) |
|---|---|---|---|---|---|
| Sudan | Bara | 1966-67 | 0.33 hr/day | n.a. | } Digerness (1977) |
| | | 1976-77 | 1-2 hr/day | n.a. | |
| Tanzania | (hills) | 1975-76 | 1.6 hr/day | n.a. | Fleuret and Fleuret (1978) |
| Kenya | n.a. | n.a. | 3-3½ hr/day | n.a. | Earthscan (1983) |
| Ghana | n.a. | n.a. | 4-5 trips/week | 2½-7 miles | DEVRES (1980) |
| *Latin America* | | | | | |
| Peru | (a) Pincos (highlands) | 1981 | 1.33 hr/day | n.a. | } Skar (1982) |
| | (b) Matapuan (highlands) | 1981 | 1.67 hr/day | n.a. | |

*Notes:* *Firewood is noted to be collected principally by women and children in all the studies listed, with the exception of Java where the labour put in is primarily by men.

†Average computed from information given in the study.

n.a. = Information not available.

to counteract this; and the present rate of reforestation is assessed by him to be less than 10 per cent of that necessary to supply the minimum needs of these countries by the year 2000 (Spears, 1978: ii, 15).

Deforestation, in turn, can have a range of devastating consequences both for the society and for the individual. At the social level these relate to the effects of soil erosion, flooding, climatic maleffects, the spread of deserts, the drying of previously perennial streams, an increased frequency of landslides in the hill areas, the rapid siltation of rivers and reservoirs, and so on (Eckholm, 1976; Digerness, 1979). As a result, there is also an adverse effect on agricultural production which has implications both for society as a whole and for those dependent on agriculture for their livelihood. By one estimate, 40 per cent of the farmers of the Third World live in valley lands and depend heavily for their irrigation water on the 'sponge effect' of forests in surrounding catchment areas (Myres, 1978: 951). The disappearance of forests produces a tendency for rain water to be released in floods during the wet-season, followed by drought in other seasons. In the forested zones of Indonesia, Malaysia and the Philippines, the green revolution is assessed to be losing its momentum because farmers can no longer find regular

*Riverbank erosion with flooding: Zambia*

*Mrs. J.H. Storrs/ICIMOD*

supplies of irrigation water for multiple rice farming (see Myres, 1978: 951).[16]

For the individual, in addition to the indirect implications of ecological destruction caused by deforestation (including the effects on farm production) are the direct ones, relating especially to the decreasing supplies of wood for fuel. Some of these noted or likely direct consequences are enumerated below.

To begin with there is a substantial increase in the time and energy spent (especially by women and children) to gather firewood. In parts of Sudan, over the past decade or so, the time taken to collect firewood has increased over four-fold (Digerness, 1977: 15). In Nepal, in the foothills of the Himalayas, journeying to gather firewood and fodder is now found to take a whole day where a generation ago the task took an hour or two (Eckholm 1975: 7). In India, in parts of Bihar where 7-8 years ago women of poor rural households could get enough wood for self-consumption and sale within a distance of 1.5 to 2 km, they now have to trek 8-10 km every day. Similarly, in some villages of Gujarat (India) the women, even after spending several hours searching, do not get enough for their needs, and have to depend increasingly on the roots of trees and on weeds and shrubs which do not provide continuous heat, thus increasing their cooking time as well (Nagbrahman and Sambrani, 1983: 36, 37). At times even straw is used. Given the already heavy working day of most rural women, any further increase in their time spent on firewood collection and cooking becomes an overwhelming burden.[17] Also, Hoskins (1983) notes how in parts of Africa, because of the extra time needed to collect firewood, daughters are now taken out of school to help their mothers. For the mothers themselves there might be a resultant economic cost in terms of employment or other income-earning opportunities foregone.

A related aspect is the greater time now spent by women in cooking because they have to adapt their cooking methods to economize on firewood. In ecological zones where wood has generally been scarce, conventional cooking practices are already such as to conserve wood at the cost of greater time and effort spent on preparing the meal. For example, in parts of the African Sahel, women light only the ends of logs and branches, placed like the spokes of a wheel, and cook with the heat generated from the ends of the spokes (Howe, 1977: 83).

Where adequate fuel is not obtainable despite the extra time and effort spent in firewood collection, there have been noticeable

*Cooking with straw in Nepal*

*Tom Learmonth/Earthscan*

changes in consumption patterns. In some areas families have had to reduce the number of meals cooked, as in Bangladesh (Hughart, 1979: 27) and the African Sahel (Floor, 1977: 6): in the latter region there has been a shift from cooking two meals to one meal a day. In other cases there is a shift to less nutritious foods: in Guatemala, to conserve firewood, families are shifting away from beans (which were a part of the staple diet) because they take too long to cook (Hoskins, 1979a: 7). In the Sahel, the diet of millet is rarely supplemented by meat, especially because a large quantum of firewood is needed in its preparation (Howe, 1977: 84); also due to fuel shortages, families in this region are noted to be shifting from millet to rice because rice takes less time to cook (Hoskins, 1979a: 7). Again, in eastern Upper Volta, attempts by the government to introduce the cultivation of soyabeans is being resisted by the women because of the longer cooking time and greater quantity of fuel that soyabeans require relative to the traditional cowpeas (Hoskins, 1979a: 7).

Necessity is also driving people in some areas to shift to food which can be eaten raw, or to eat partially cooked food (which could be toxic), or to eat cold leftovers (with the danger of food rotting in a tropical climate), etc. (Hoskins, 1979: 7).* All this increases the vulnerability to ill-health and infection. In some regions, adverse nutritional consequences are also found to result from the trade-off between time spent in gathering fuel and that spent in cooking. In Peru, for instance, in areas where firewood is scarce and its collection requires women to travel far from home, taking several hours, much less time can be spent by them in cooking: this is noted to adversely affect the quality of the family's diet (Skar, 1982).

Increased deforestation also adversely affects the availability of forest produce other than firewood, which is gathered by the local people for their own needs and often for sale as well. This includes not only fruit and fodder, but also herbs and plants for local medicines (Hoskins, 1982), turpentine and resin (Guha, 1983), certain kinds of flowers, such as *mahua* in India, which provide oil, liquor and cattle

---

*The complementarity of food and fuel is also reflected in the following proverbs (quoted in Smil and Knowland, 1980: 7, 12):
'It costs as much to heat a pot as to fill it' (old African proverb).
'It's not what's in the pot but what's under it that worries you' (old Chinese saying).

feed (Chand and Bezboruah, 1980; Narayan, 1982), the seeds and pods of various trees such as sal and *neem* which provide oil, the twigs of the *neem* which are used to brush teeth in India, the leaves of the sal which are used as disposable plates (Bhaduri and Surin, 1980), the leaves of the *tendu* from which *biri* (a type of Indian cigarette) is rolled (Guha, 1983), various types of fibre and floss, tans and dyes, and so on.[18] By one estimate some 30 million people in India (mostly tribals and forest dwellers) depend on such minor forest produce for some part of their livelihood (Kulkarni, 1983: 194); and they are getting increasingly marginalised.

Additionally, deforestation has been noted to lower the water table, making it much more difficult and time consuming for women to obtain water, especially during the dry season. This is noted for instance in Senegal (Hoskins, 1983), in India (Bahuguna, 1984), and in Nepal (Dogra, 1984). As a woman of the Uttarkhand hills in India puts it:

> 'When we were young, we used to go to the forest early in the morning without eating anything. There we would eat plenty of berries and wild fruits...drink the cold sweet (water) of the *Banj* (oak) roots.... In a short while we would gather all the fodder and firewood we needed, rest under the shade of some huge tree and then go home. Now, with the going of the trees, everything else has gone too.'[19]

In some areas, such shortages are found to be linked crucially to life and death questions. In the above mentioned region of India, for example, a woman grass-roots activist has noted several cases during the past three years of young women committing suicide because of the growing hardship of their lives with ecological deterioration. Their inability to obtain adequate quantities of water, fodder and fuel leads to scoldings by their mothers-in-law (in whose youth forests were plentiful), and soil erosion has made it much more difficult for the women to produce enough grain for subsistence in a region of high male out-migration. In one year seven such cases of suicide were observed—four of them in a single village where shortages are especially acute (Bahuguna, 1984).

Another manifestation of shortages is the significant rise in the price of firewood along with its increasing commercialisation. While this price rise is observed mainly in the urban areas, with prices in

*Searching for water in dry riverbeds: Orissa, India (above)
and Ethiopia (below)*

many Third World cities having more than doubled over the past decade,[20] it also adversely affects those among the rural population who have to depend on firewood purchase. These are mostly the poor households with few personal resources of land or cattle to provide fuel.[21] Also, as firewood gets commercialised, it will become increasingly difficult for such households to gather it free or at minimal cost.

Further, in general, there is noted to be a shift from firewood to other fuels such as cattle dung, maize stalks, etc. The use of cattle dung, in particular, has a high opportunity cost[22] in terms of the loss of agricultural output that could have been produced had the dung been used as manure instead. By one rough estimate, for every ton of cattle dung burnt there is a loss of 50 kg of foodgrains,[23] and given the further estimate that in the Third World about 400 million tons net weight of cattle dung is burnt every year, this could amount to a loss of 20 million tons of grain annually (Spears, 1978: 4, 5).[24]

In short, the signs of an existing woodfuel crisis are undoubtedly apparent in many areas, and those of an impending one in many others.

*Substituting dung-jute sticks for fuelwood: Bangladesh*

Marty Chen

## 4. Whose Crisis Is It?

Clearly, the shortage of woodfuel is not experienced as a hardship by all in equal measure. There is a socio-economic dimension to the severity of its impact as well as a regional one.

The socio-economic dimension relates to the fact that the implications are likely to be more serious for the poorer (especially landless) households who usually depend for their firewood supply, particularly in the rural areas, on what they can gather freely, and who are often not in a position to substitute firewood with alternative fuels that require cash expenditure. Hence, if the poor cannot gather firewood their nutrition suffers, while the economically well-off either purchase firewood—if available at a price—or substitute it with other purchased fuels. Also, as firewood gets scarcer and increasingly monetized, those who cannot enter the market will tend to be squeezed out. For such households, there can be a trade-off between buying fuel or food. In a Bangladeshi village, for example, landless Hindus who are both economically and socially the worst off in the village, have to buy firewood during the monsoon (when crop residues etc. are not available) at the cost of buying food (Briscoe, 1979: 627, 632).[25] This also highlights the seasonal dimension of the problem.

Women in poor households bear an additional burden. First, as the main gatherers of fuel it is primarily *their* time and effort that is extended with shortages. Second, they face more severe nutritional consequences from such shortages than men because of the biases against them in the distribution of food within the family—as has been noted in many parts of Africa and Asia.[26] The extra energy expended by women of poor households to collect fuel is also unlikely to be made up in most cases by the required higher consumption of food.

The regional dimension of the crisis relates primarily to three aspects: its country-specificity, its zone-specificity and its rural-urban implications.

To begin with it is essentially a Third World crisis—wood provides less than 1 per cent of the energy in most First World countries (Earl, 1975: 10). Further, within the Third World it is (or is likely to be) felt more severely in some countries than others. For example, by one estimate[27] in as many as 14 countries in Africa (13 located south of the Sahara), 4 in Asia and 2 in Latin America and the Carribbean, current wood use (for fuel and non-fuel purposes) is either very close to or higher than estimated sustainable forest yield[28]—and the gap is expected to widen over the next decade.

Table 6 gives an idea of the number of people in acute scarcity, deficit, or prospective deficit situations in 1980 and their likely

## Table 6: Populations involved in fuelwood* deficit situations
### (in millions)

| Region | 1980 Acute scarcity | | 1980 Deficit | | 1980 Prospective deficit | | 2000 Acute scarcity or deficit | |
|---|---|---|---|---|---|---|---|---|
| | Total | Rural | Total | Rural | Total | Rural | Total | Rural |
| Africa | 55 | 49 | 146 | 131 | 112 | 102 | 535 | 464 |
| Near East and North Africa | — | — | 104 | 69 | — | — | 268 | 158 |
| Asia & Pacific | 31 | 29 | 832 | 710 | 161 | 148 | 1671 | 1434 |
| Latin America | 26 | 18 | 201 | 143 | 50 | 30 | 512 | 342 |
| Total | 112 | 96 | 1283 | 1053 | 323 | 280 | 2986 | 2398 |

*Note:* *Wood for charcoal is included here. Hence strictly in terms of the definition used in this book, these figures would relate to 'woodfuel' not just to 'fuelwood'.

*Source:* FAO (1981): *Map of the Fuelwood Situation in Developing Countries—Explanatory Note*, FAO, Rome.

position in 2000 AD. By these estimates, as many as 1.4 billion people are today in acute scarcity or deficit; of these 82.4 per cent are in the rural areas. By the year 2000 the number in acute scarcity or deficit will together be 3.0 billion, of which the bulk will be in the rural areas.

Further, the crisis is more severely felt in some ecological zones than others. For example, the social consequences of ecological maleffects would tend to be more severe in arid and semi-arid regions such as in the African Sahel or the Indian State of Rajasthan. The private consequences too are likely to be more serious here, since the absolute availability of wood is lower in the arid zones than in the tropical ones, and locational availability, as noted, is specially important for the poor.

National averages of likely demand-supply balances noted in the Hughart estimate above, do not bring out this inter-regional dimension. Hence even a country where the overall demand-supply balance does not reflect a shortage could be experiencing severe region-specific problems, due to differential cross-country deforestation and/or the geographical inaccessibility of existing forests.

*Encroaching desert: Rajasthan, India*

Additionally, the crisis has different rural-urban implications. On the one hand, it could be seen as being more severely felt in the rural areas because, in general, the availability of alternative energy sources is much higher in the urban than in the rural areas.[29] Also, as urban demands increase and firewood sales become increasingly profitable, the quantities flowing from the rural to the urban areas will rise. Already in many cities this flow is significant: for example, firewood into Delhi is noted to come largely from the forests of Madhya Pradesh, 700 km away, where an estimated 6 ha must be clear-felled daily to meet the city's requirements (Agarwal, 1983b: 19). However, on the other hand, the fuel (whether woodfuel or other) available in the urban areas in largely monetised, so that for the urban *poor* who cannot afford cash expenditure the problem could be much more serious than for the rural poor, who usually would be able to forage some crop residues or cattle dung. In terms of the *number* of people involved, however, the problem of woodfuel shortages is essentially a rural one.

In overall terms, therefore, it is clear that the consequences of woodfuel shortages are likely to be more serious for the members of poor households (most of whom are rural-based), especially in the arid regions of Third World countries. However, it is by no means obvious that it is the gathering of wood for fuel by the poor that has caused the crisis.

## 5. Causes of the Crisis

In considering the causes of the woodfuel crisis a complex set of factors is encountered. Impinging on the issue are three closely interrelated aspects: one, of the absolute availability of wood for all uses (which depends on the tree resources of the country, barring imports); two, of the availability of wood for fuel (which depends on the distribution of available wood supplies between different uses); and three, of the availability of woodfuel to the poor (which depends on the distribution of available woodfuel supplies between people). As elaborated below, woodfuel shortages being faced by the poor today emerge from the particular uses to which forest land and wood resources have been put over the years by specific classes of people, and cannot be traced in any straightforward way to the gathering of firewood by the poor for their domestic use.

To begin with, the depletion of forests and the associated reduction in the absolute availability of wood for all uses is related to the exploitation of land for various purposes. It is noteworthy that large tracts of forest land have been and continue to be cleared for crop cultivation, pastures, plantations, industrial development, etc. Rising population pressures, State policies and private profit considerations would all be contributory factors impinging on this land clearance. According to one estimate for India, of the total forest area officially recorded as deforested for different purposes between 1951-52 and 1975-76, as much as 60.6 per cent was for agricultural uses (see Table 7). Some authors maintain that one of the significant causes of deforestation, not only in India but also in other parts of Asia, is shifting agriculture.[30] Traditionally, under this system, some forest land used to be cleared, cultivated for a limited period, then left fallow to regain its fertility while the cultivator moved into another stretch of the forest. The authors note that several thousand hectares are still under this system of farming in the hilly parts of the Assam, Tripura, Manipur and Orissa States of India. In Africa, 'slash-and-

**Table 7: Area deforested in India for different purposes**

(1951-52 to 1975-76)

| Purpose | Area deforested | |
|---|---|---|
| | Million hectares | Percentage |
| Agricultural uses | 2.51 | 60.6 |
| River valley projects | 0.48 | 11.6 |
| Establishment of industries | 0.13 | 3.1 |
| Road construction | 0.06 | 1.5 |
| Miscellaneous | 0.96 | 23.2 |
| Total | 4.14 | 100.0 |

*Note*: The figures relate only to notified areas which have been transferred officially, and do not include illegal encroachments.

*Sources*: Swaminathan (1980): 'Indian Forests at the Crossroads', in *Community Forestry and People's Participation: Seminar Report*, Ranchi Consortium for Community Forestry, November 20-22.

burn' agriculture has traditionally been (and in some places continues to be) a common practice in most tribal communities. While earlier, the ecological balance was maintained through sufficiently long fallows to allow tree growth, now with the increasing number of dependents on the land and the greater competition for land for other (e.g. industrial/commercial) uses, this is becoming less and less possible.

Forest land has also been systematically cleared to establish plantations or pastures in many parts of the Third World. In India, under colonial rule, vast tracts of mountain forest were given away to individuals for setting up tea and coffee plantations, in addition to encouragement being given to forest clearance for extending crop cultivation, in order to augment the government's land revenues (Guha, 1983: 1883, 1893). In Latin America, again, the conversion of private forests into pastures is noted to have contributed significantly to deforestation. In Brazil and Venezuela, for example, large tracts of forest were so converted by cattle ranchers eager to make a profit on beef (Eckholm, 1979: 17). The clearing of trees for mining and quarrying activities (to obtain coal, copper, limestone, etc.) is another important source of deforestation, as in India. State policies, as well as the control exercised by the economically privileged sections

*Smitu Kothari*

*Deforestation and soil erosion due to mining: Madhya Pradesh, India.*

of the population on valuable natural resources, can thus be seen to impinge significantly on the pattern of use of these resources.

Additionally, in some areas forest devastation has been compounded by the foraging of shrubs and leaves to feed cattle. According to an FAO estimate for Nepal, one buffalo eats 7 tons fresh weight of leaves per year, and a cow 2.5 tons; all this leafy material comes from the forests. Given the 14 million farm animals in 1969, which exceeded the 11.2 million human population in that year, it is not surprising that domestic animals in Nepal are estimated to annually consume twice the equivalent of forest biomass used as firewood (Shephard, 1979: 81, 82).

Next, it is important to take into account that available resources of wood from forests have a variety of alternative uses. For example, the commercial exploitation of forests to supply timber for building purposes, for direct export, or to supply wood as raw material to industries such as paper and rayon manufacture, has gone on historically and continues today. Some authors maintain that the cutting of trees for timber is one of the principal causes of deforestation in the Third World.[31] Historically, such exploitation was carried out virtually indiscriminately under colonial rule in many Third World countries. India provides a graphic illustrative example. Guha (1983) in a detailed review of forest policy in India during colonial and post-colonial rule quotes one British scholar as noting that British rule saw a 'fierce onslaught on India's forests' especially for the expansion of railways in the mid-19th century.[32] To build railway sleepers, vast tracts of the more accessible forests were 'felled in even to desolation' in the Garhwal and Kumaon hills, under the charge of Indian and European private contractors.[33] Often no supervision was exercised on the activities of the contractors, leading to enormous wastage as 'thousands of trees were felled which were never removed, nor was their removal possible'.[34] Age-old hardwood deodar forests were cut down in the north-western Himalayas: from this region alone, between 1869-1885, wood for building 650,000 deodar sleepers was supplied. The use of local timber as fuel for the railways also continued upto the 1880s causing considerable deforestation in the north-western provinces.

Guha notes that the colonial exploitation of forests was further intensified during the two World Wars, when there was an estimated increase of 65 per cent over the pre-war out-turn of timber for building ships, bridges, etc. and in little over one year (April 1917 to

October 1918) over 228 thousand tons of timber are estimated to have been supplied for such uses in addition to that supplied for railway sleepers. Also, about 1.7 million m³ of timber (mostly teak) was annually exported between 1914-1919, and 'fellings and sawings were pushed into the remotest forests in the Himalayas, and into the densest jungles of the Western Ghats'.[35] The revenue of the forest department in 1924-25 was Rs.56.7 million, giving a surplus of Rs.48.9 million. Again in the post-colonial period (post-1947), by Guha's analysis, commercial and industrial interests have dictated forest policies, with a noteworthy growth in demand in the paper industry: over the period 1966-77 the production of pulpwood has shown a 400 per cent increase, although it still represents (by official estimates) only a small percentage of total roundwood production in the country.

Nepal's historical experience appears to have been similar in several respects. Bajracharya's (1983a) analysis indicates that from the 18th century up to 1950 there was a large-scale cutting down of trees with little adherence to scientific principles of forest management—the beneficiaries of the deforestation being basically the elite. Also, even though Nepal was not directly under colonial

*Commercial exploitation of scarce forest resources: India.*

Smitu Kothari

rule, its forest policy after 1925 was strongly influenced by the British Indian Forest Services, whose advice the Nepalese government sought. During World War I, when demand in India for sal wood was high, the British are noted to have advised Nepal to undertake intensive felling in the West Nepal forests and to clear forest land in some areas for crop cultivation. When demand fell sharply in the 1930s, large quantities of already-felled timber which could not be exported were laid waste. While the post-1950 government policies are said to be designed for forest conservation, Bajracharya's (1983a) assessment of these policies is that they are not 'people oriented' and do not take account of the needs of the local communities.

The outright export of timber is also important to take into account when considering the causes of deforestation. This export has taken place not only under colonial regimes, as in India, but continues under national governments as well. For example, the depletion of tropical hardwood forests in S.E. Asian countries, such as Indonesia, the Philippines and Malaysia, is attributed by some observers to the export of logs and processed timber to developed and rich developing countries (Myres, 1978).[36]

This trade has assumed major proportions in recent years. In 1978, for example, by one estimate, as much as 55.1 per cent of total logs produced in southeast Asia were exported, and in the same year 63.5 per cent of the production of logs plus processed wood in southeast Asia, west Asia and the Pacific taken together was exported (Leslie, 1980: 5). As the author puts it: 'Exploitation, for once, is the right word to use. Not even the most generous assessment could describe what is happening to the major part of the southeast Asian tropical forests in any other terms. Forests, as a whole, are simply being mined, taking out the easiest to get—and the most highly priced trees—without any real concern for what happens afterwards' (Leslie, 1980: 3). The felling is carried out by indigenous companies or transnational corporations to whom tracts of government or public or community-owned forest land are allocated by the State.

A number of Third World countries, including Kenya, Sri Lanka, Somalia, Thailand, Mexico and Tunisia, have also been exporting charcoal.[37] While most of these countries are believed to have recently stopped such exports,[38] Thailand and Mexico continue to export increasing quantities, and Indonesia is a new and significant entrant in the trade. Ironically, the bulk of these exports are to oil-rich Third World countries, especially in the Persian Gulf, where charcoal is preferred as a cooking fuel; some amount also goes to Japan and Western Europe (Knowland and Ulinski, 1979: 13).

The use of wood as a fuel is clearly also of considerable importance but in the absence of systematic macro-estimates of wood consumption by end use, existing assessments of the proportion used as fuel are at best rough and ready. By one estimate (FAO, 1978a: 90-91) 84 per cent of all wood consumed annually in the Third World today is so used, but limitations of the FAO data on which this estimate is based would have affected its accuracy.[39] In any case, even if we take this assessment as indicative, this in itself cannot establish the gathering of wood for domestic use as a significant cause of *deforestation*, especially not *vis-a-vis* the rural areas. This is because firstly a not insignificant percentage (estimates vary between 11-25 per cent[40]) of woodfuel is used in small scale and cottage industries such as brick making,[41] pottery,[42] blacksmithy, bakeries, the curing of tobacco[43] and leather, fish drying and curing, etc. Second, survey evidence indicates that the firewood consumed in rural homes is often in the form of twigs and branches, the gathering of which would cause no deforestation relative to the actual cutting of trees which would.

For instance, according to measurements made by the Karnataka Institute of Science (India) in a Karnataka village, 90 per cent of the village fuelwood supply comes from twigs and small branches (Makhijani, 1979: 24). Also, in the same village, 96 per cent of wood-gatherers use mainly twigs for fuel. Those purchasing firewood are noted to buy either branches or roots, not logs (Ravindranath, *et al.*, 1978). Likewise, in a recent study of household fuel use in South Korea, twigs were found to account for 73.4 per cent of total fuelwood consumption by weight.[44] Briscoe's study in Bangladesh again indicates that much of the wood gathered in the village (especially by the landless) is in the form of twigs and fallen branches. Studies such as these, which provide information on the form in which fuelwood is consumed domestically, although unfortunately few in number, are nevertheless strongly indicative.

To sum up this argument, woodfuel shortages being experienced today are closely related to the form and degree of forest exploitation that has taken place over several decades *in the past* (as a result of specific State policies, private profit considerations, demographic pressures, etc.). This exploitation has taken a variety of forms, such as the clearing of forest land for agriculture, plantations, pastures and other uses; the commercial logging of trees for timber either for building purposes or for exports; the use of wood as an industrial raw material and as a fuel in small-scale and cottage industries, and so on, and is not necessarily associated with the use of wood as a domestic fuel. Even today, the deforestation that is occurring cannot be attributed straightforwardly to the gathering of wood by poor rural households for their domestic use, since a good deal is likely to be as twigs, etc. It is noteworthy that the use of wood for almost all purposes other than as a domestic fuel in the rural areas, requires the cutting down of trees.

Even to the extent that in some areas trees are being cut or the barks of trees being stripped off by the poor to obtain fuel, it must be seen as a *symptom* of the crisis—a reaction of the people who are the severest hit by it but who cannot be held responsible for having *caused* the crisis. As one landless woodcutter in the Sudan put it: 'We take trees belonging to other people. We cut them when they are too young. We never pay royalty. . . . We are in a miserable state after our animals starved to death during the drought. We must live for something. What else can we do?' (Digerness, 1977: 12). What indeed! With no land and no employment, cutting trees for fuel and producing charcoal for sale is their only means of survival in this context.

This last observation brings us to the major aspect relating to the experience of woodfuel shortages as a 'crisis', namely the maldistribution of material resources between different classes and social groups within a country, which affects both the absolute availability of wood and its relative distribution between uses and users. For example, the use to which land is put, and the effect this has on deforestation, is related to the unequal distribution in the ownership of and access to land among people. Thus, on the one hand, large tracts of forests on both private and public land are often cut down for the private gain of a few, and on the other hand, an increasing number are forced to survive on diminishing tracts of 'common' forest, which they are then blamed for destroying. As one recent study in the villages of Rajasthan and Madhya Pradesh (India) noted, common property resources are crucial for the survival needs of those with little or no land: 86 per cent of agricultural labourers and small farmers in the surveyed Rajasthan villages and 98 per cent in the Madhya Pradesh villages depend on common land for fuel while none of the large farmers in either region do so. As more and more of such land is being brought under private ownership (with much of it going to the larger farmers) the poorest are forced to depend on shrinking areas of common land (Jodha, 1983: 8, 9).

Here the issue is not one of 'weaning tribal people from their age-old practice of shifting cultivation' by 'careful demonstration' (of improved methods) and 'persuasive measures' as some observers emphasize,[45] but rather of radical agrarian reform for bringing about a more egalitarian pattern of land ownership and control, and of ensuring alternative, viable means of livelihood for the poor.

Again, the persistence of the illegal cutting down of government-owned forests by local timber merchants who make large profits on public resources (as observed in parts of south Asia) is closely linked to the economic and political power that these men command. This enables them not only to control the local villagers but also to ensure the allegiance of the local government officers.[46] The landless, in contrast, are liable to face persecution if caught collecting even small amounts of firewood without a permit (BRAC, 1980, Swaminathan, 1982a).

In other words, the maldistribution of material resources, by determining who has access to available wood supplies, also affects the *relative* availability of woodfuel to different sections of the population, and hence the relative cost of the shortages to different people.

Further, the severity of the woodfuel crisis is accentuated by the high and rising prices of alternative fuels such as kerosene. This, in turn, is closely related to the dynamics of supply-demand that operate between the oil-exporting countries and the developed oil-importing ones.[47] The unequal access of non-oil-rich Third World countries *vis-a-vis* the rest of the world constitutes the international dimension of a country's internal crisis.[48]

In the light of these factors a search for solutions to alleviate the problem of woodfuel shortages experienced by the poor becomes a complex one, as will be discussed further below.


## 6. The Search for Solutions

In technical terms, the 'solutions' most commonly outlined in current literature concern the promotion of two categories of innovations:

(i) Woodfuel-related innovations, especially improved wood burning stoves and tree-planting schemes—the former representing a way of increasing the efficiency with which existing supplies of firewood can be used and thus of increasing the effective fuel energy available to the household, and the latter a way of increasing the quantum of existing supplies of firewood. Both these can be seen as having the potential for making a direct impact on the problem of woodfuel shortages faced in the rural areas. [Improved equipment for converting wood to charcoal or other secondary fuels is also sometimes emphasized but since charcoal (as noted) is largely monetized, is used primarily in the urban areas, and constitutes only a small percentage of total woodfuel energy used in most Third World countries, this cannot be seen as having the same importance in the present context as stoves or tree-planting[49]].

(ii) Non-wood-related (renewable) energy technology such as solar cookers, biogas plants, etc.

However, when presented as a kind of 'checklist' of items (as they often are), such 'solutions' fail to come to terms with the central question: how can the successful adoption of these innovations be ensured?

As will be seen from the discussions in Chapters III to V and the Appendix the problem of the diffusion of innovations in the rural areas is not always an easy one to solve. A complex set of economic, social and political factors impinge on both the appropriateness of particular innovations, and the feasibility of their adoption. For example, it will be shown that even though *conceptually* the process of innovation-generation (the technical aspect) can be separated from that of diffusion, such a separation is often not desirable in practice. *In practice* the degree to which the user is involved in innovation-generation itself, can be significant in determining whether or not the innovation is adopted, or continues to be used when adopted. This is not merely because the user would then develop a personal interest in the success of the experiment, but also because the innovation is then likely to be more closely moulded to the user's needs. These needs, which are often culturally and socially determined (as, for example, the method of cooking), can rarely all be anticipated or taken into account in the laboratory.

The practical implications of bringing about user-involvement, however, are likely to be wider than apparent at first glance. For instance, where user-involvement is a necessary condition for successful diffusion, further questions arise: to what extent can such involvement, which would require a close and harmonious interaction between the scientific community (the scientists, technicians, etc.), the bureaucracy (the village extension workers), and the village community (often the poor peasant households) be ensured, given the material and social (status) hierarchies and inequalities found between these groups in most Third World countries, and which strongly influence the attitudes of these groups and their ways of relating with one another. Even interpersonal interaction among those comprising the village community tends to be affected by differences of class, caste, gender, etc. and the possible conflict of interests to which these give rise.

Similarly, in reforestation schemes, a consideration of issues of land ownership and land tenure patterns may be unavoidable. Tree planting involves choices about land use (whether privately or communally owned); and the pattern of land ownership and control can be significant in determining how the costs and benefits of the scheme are distributed between members of the community (which in turn is likely to affect the willingness of different members to participate in the programmes).

The success of the technical solutions is thus likely to depend on the degree to which these realities of social structure are taken into account and dealt with.

It is necessary to mention here that the promotion of woodfuel-related innovations would constitute only a part of the answer to the problem of woodfuel shortages faced by the poor. An issue of equal importance in the search for an *effective* solution is that of the *distribution between uses and users*, both of wood and of non-wood-related energy sources. As already noted, a woodfuel crisis makes itself felt not merely because there is some overall supply-demand imbalance in wood availability in the country, but also because what is available is unequally distributed between people and inappropriately distributed between uses[50] (in addition to the inappropriate distribution of non-wood energy resources between uses and users).

This raises a number of basic questions. For example, if the distribution of wood between different uses is important, the appropriateness of a charcoal export policy in countries with severe woodfuel shortages, or of licensing policies that lead to the indiscriminate commercial exploitation of forests for timber by private contractors, may be queried. At a broad level, the choice between different uses of available energy (woodfuel or other) cannot be separated from the question: what products are being produced in the country, and what technologies—with their associated implications for sources and levels of energy use—are being used to produce them? In this sense, then, the issue is inseparable from that of the overall development strategy being pursued by the country.

Both aspects—the distribution of energy between *uses* and the distribution of energy between *users* (like the issue of woodfuel innovation diffusion), ultimately impinge on central aspects of political economy—on the distribution of material wealth and political power between groups within the country, which determines *who gets* access to *how much* of a scarce resource, and *for what purpose*. The international dimension of this issue likewise relates to the distribution of available energy resources between countries.

Some of these questions—such as that of the distribution of wood and non-wood energy between uses—go beyond the scope of the present study, the focus of which is on the issue of woodfuel innovation diffusion. However, even in this context, as noted, a consideration of issues relating to the distribution of material assets and available resources between people is likely to be of critical importance.

## Notes

1. A collection of these songs is to be found in *Chipko Geet* (1979).
2. The cost in foreign exchange for oil-importing Third World countries has increased sharply over the past several years (see Prasad, 1979: 388, for India; and Tuschak, 1979: 2, for Kenya).
3. For a useful discussion of the problems in a number of available macro- and micro-estimates see Hughart (1979: 34-35); Gamser (1979: 5-11); Earl (1975: 58).
4. See Makhijani (1979); Chung (1979); and Briscoe (1979). This aspect is discussed in more detail further on.
5. Earl notes that the FAO *Yearbook* figure for recorded removals in 1970 is shown by survey to be about 84 per cent of the actual fuelwood consumption in that year.
6. There is some recognition in recent FAO literature of the likely underestimation of fuelwood consumption in FAO's annually published figures. There has also been some attempt to adjust macro-statistics on the basis of survey evidence. But the problem has no easy solution. To begin with, surveys of household fuelwood consumption do not exist in most countries, and many of the surveys that do exist are of questionable accuracy (see especially, Barnett, 1982; and Gamser, 1979).

Almost all of the few carefully done surveys that have emerged over the past few years relate either to a single village or to a group of villages in one region. These, while extremely useful in providing significant pointers, yet present problems for generalising about magnitudes of consumption for the country as a whole, since there tend to be regional variations in fuel availability and use. Hence adjustments made in macro-statistics on the basis of existing surveys (as attempted for instance in FAO, 1978a: 90) cannot but have an element of arbitrariness.

7. The total energy figures used in Table 1 relate to energy used for all purposes, in all sectors, in the rural and urban areas combined. If we were to consider energy use only in the household sector and further only in the rural areas, the importance of woodfuel would be greater still. In India, for instance, while woodfuel is noted to provide only about one-third of total energy used, it is estimated to provide over two-thirds of energy consumed in the rural household sector (see Government of India, 1982: 6).

8. See Openshaw (1978), reproduced in DEVRES (1980: 45).

9. Firewood, agricultural residues, and animal wastes are usually non-monetised; hence in many studies these three sources are termed 'non-commercial', though strictly speaking this categorization would not be applicable in all situations.

10. A summary table presented by Hughart (1979: 63) based on survey data of household fuel use indicates that over 95 per cent of the populations of Gambia, Nepal, Sudan, Tanzania, and Thailand depend solely on firewood and charcoal.

11. Biogas is a mixture of methane (60-70 per cent), carbon dioxide and small quantities of hydrogen sulphide, nitrogen, hydrogen and carbon monoxide. It is inflammable and is produced when organic matter, such as human and animal waste, garbage, twigs, grass, leaves, agricultural wastes etc., is fermented under airtight conditions at suitable temperatures, and with an appropriate amount of water and level of acidity in the environment. Slurry which can be used as fertilizer is a by-product (for further technical details see Van Buren, 1979).

12. See Barnett *et al.* (1978: 92 to 95); Moulik (1978); and Reddy and Prasad (1977).

13. 1 US dollar = Rs.12.

14. Arnold and Jongma (1977); Arnold (1978); Floor (1977); Nkoniki (1978); Openshaw and Morris (1979); Siwatibau (1978); Spears (1978); Hoffman (1980).

15. The illegal nature of their activities leaves them open to exploitation both by petty State officials (forest guards, railway staff, etc.) who extract a payment per bundle and, in some villages, by village headmen who take a tax (Bhaduri and Surin, 1980).

16. For further documentation on this aspect especially as relating to Latin America and Africa see Eckholm (1976: 91-96).

17. Rural women, especially of small peasant households, are usually noted to work an average of 9 to 10 hours per day in farmwork plus domestic chores; and in some areas they work even 15 hours per day. Typically the time they put in is more than that put in by men of the same class of household (for a survey of material on this see Agarwal, 1985).

18. For a detailed breakdown for India of the different types and quantities of minor forest produce collected during specified seasons of the year see Guleria and Gupta (1982).

19. Quoted in Bahuguna (1984: 132).

20. See, e.g. Eckholm (1980); DEVRES (1980); Agarwal and Bhatt (1983).
21. At the same time, such price increases will not necessarily benefit those among the poor who sell firewood for a livelihood. In the Indian context, for instance, any such gains are likely to pass onto those noted to extract payment from the headloaders for allowing them to collect and sell the wood.
22. The costs of the opportunities foregone in terms of the alternative uses of a scarce resource.
23. Such estimates are of course rough and ready and depend on specific assumptions regarding the manure response rates of particular crops, under particular soil and climatic conditions. However, they help to illustrate the basic point.
24. This could also in some sense be viewed as a 'social' cost and not merely as a private one.
25. Also see Cecelski (1984: 48-49).
26. See, e.g. Schofield (1979), Longhurst (1980); Agarwal (1985).
27. See Hughart (1979: 79-98). The estimates are somewhat rough and ready but still useful in providing a broad idea of the geographical spread of the problem.
28. These countries are the following: Burundi, Ethiopia, Ghana, Guinea, Kenya, Mauritius, Nigeria, Rwanda, Sierra Leone, Swaziland, South Africa, Tunisia, Uganda, Upper Volta, Afghanistan, Bangladesh, India, Pakistan, El Salvador, and Haiti.
29. For one of the best-known expositions of the 'urban-bias' thesis, see Lipton (1977).
30. See e.g. Novak and Polycarpau (1969: 21, 22).
31. Novak and Polycarpau (1969: 22, 23); and Makhijani (1979: 24).
32. Smythies (1925) quoted in Guha (1983: 1883).
33. Pearson (1869) quoted in Guha (1983: 1884).
34. *Ibid.*
35. Government of India (1948) quoted in Guha (1983: 1887).
36. Indonesia is noted to have leased out most of its forest estate to international timber corporations (Myres, 1978: 951).
37. See Knowland and Ulinski (1979: Appendix 2); and Uhart (1975: 9). It could be argued that the export of charcoal would enable the import of oil-based fuels. However, there is no certainty that the foreign exchange earned from the exports would be used for this purpose or, if it were, that the oil would be made available for domestic use rather than for industrial use. Also, such fuel would be available only at a price, which would not necessarily help the poor, for whom available supplies of firewood may have been reduced by the export of charcoal.
38. Illegal, unrecorded exports may however still be continuing.
39. This estimate relates to annual woodfuel production (published figures adjusted for survey evidence) expressed as a percentage of total annual roundwood production. Production has been assumed as equal to consumption. Roundwood is defined as all wood in the natural state as felled or otherwise harvested. Problems with FAO's woodfuel production (consumption) figures, even if certain adjustments are made, have already been noted. In the case of roundwood, FAO statistics for some countries are based exclusively on recorded volumes, and for other countries on recorded volumes as well as FAO's own estimates of unrecorded volumes. In the case of most countries such unrecorded

volumes (both for domestic and other uses) can only be guessed at, and may be quite large; and as elaborated earlier, any adjustments made on the basis of survey data (which itself is available for only some countries) are also problematic. Hence, even at best the noted estimate of the percentage of wood used as fuel is likely to reflect a high degree of approximation.

40. See Openshaw (1978: 41).

41. In Tanzania, one brick and kiln factory is noted to annually use 20,000 m³ of fuelwood, obtained mainly from the national forests (DEVRES, 1980: 50).

42. In Pura (India) four pottery establishments along with a coffee shop consume 8.9 tons of firewood annually (Ravindranath, *et al.*, 1978).

43. In Janakpur (Nepal) an estimated 76,000 tons of fuelwood is consumed annually to heat 450 tobacco barns (see Zummer-Linder, 1976: 8).

44. Computed from Chung (1979: 23).

45. For instance, Novak and Polycarpau (1969: 22).

46. See BRAC (1980) for Bangladesh; and Nath (1968) for India.

47. By some estimates the high demand for fuel in many developed countries could be reduced considerably if measures were taken to eliminate wasteful use. Goldemberg (1978: 31) quotes some recent research which indicates that approximately 40 per cent of the fuel spent in USA could be saved by conservation measures based on currently available technology, without affecting people's lifestyles, and only requiring improvements in the efficiency of existing energy consuming systems.

48. For an interesting and forceful exposition on the politics behind the international division of available energy sources see Tanzer (1979).

49. As further elaborated in the appendix, improved equipment for converting wood to charcoal would basically have an indirect impact by increasing the income of those of the rural poor who produce charcoal for sale.

50. This also affects the 'efficiency' with which wood (or any other energy resource) is used, since the concept of 'efficiency' must take account too of the allocation of wood between different uses, and not merely of how well it is used for a given purpose.

# CHAPTER II

# WOODFUEL INNOVATIONS: SOME TECHNICAL ASPECTS

*. . . cooks received a larger total dose (of pollutants) than residents of the dirtiest urban environments. . . . The women cooks are inhaling as much benzo(a) pyrene as if they smoked 20 packs of cigarettes per day.*

—Smith, *et al.* (1983)[1]

What are the characteristics of woodfuel innovations? In what ways do they represent improvements? To provide a background this chapter attempts to summarize some of the technical aspects relating to these innovations; but it is not meant to be a comprehensive survey of the technical literature on the subject. The innovations fall under three heads: improved wood-burning stoves, tree-planting schemes and improved wood-conversion hardware. The first represents a way of increasing the efficiency with which a given supply of firewood is used; the second of increasing the available firewood supply and the third of increasing the efficiency with which wood is converted to charcoal and other secondary fuels. These in themselves do not constitute 'options' in the sense of being alternatives or substitutes since, in most instances, the promotion of both the first two and sometimes also the third, would need to be part of a 'solution'. What would be optional, however, is the relative emphasis placed on the three in the 'mix' (which would be country and region-specific) and the choice of alternatives within each. Let us consider them in turn.

## 1. Improved Wood-Burning Stoves

Attempts to develop improved wood-burning stoves have been ongoing in several countries in recent years as, for instance, in India,

Guatemala, Indonesia, Nepal and so on.[2] In some countries these attempts date back over three decades. In India, for example, the Magan *chulha* (stove) was developed in 1947 and the HERL *chulha*[3] in 1953, followed by other designs such as the South Junagadh *chulha* and the Nada *chulha*. In Indonesia, again, the Singer stove was developed in 1961 (see Singer, 1961) and today a range of stoves based on the original design are being promoted. In Central America, the Lorena stove was first developed in Guatemala and modified versions of it (such as the Ban-ak-Suuf stove) are now being promoted in several African countries including Senegal and Upper Volta. The materials used vary: some are built entirely of local mud and clay, others use some metal or iron pre-fabricated parts (especially chimneys), or ceramic inserts, etc.

Improvements in the design of wood-burning stoves are aimed primarily at increasing the efficiency with which wood is burned. In practical terms, the most important aspect of this is reduction in the amount of firewood needed by the user to cook the household's everyday meal. In addition, the attempt is to improve the conditions under which cooking is done, especially by eliminating smoke.

While simple measures such as drying the firewood before burning can considerably raise its calorific value and so increase efficiency, improvements in stove and/or pot design (and in the skill of the

*Cooking with twigs on a smoky stove: India*

*Hindustan Times photo library*

person operating the stove) are necessary for more substantive efficiency increases. The idea is to improve the heating process which can be seen as involving three stages:[4]

— efficient generation of heat through the efficient combustion of wood;

— efficient transmission of the heat to the heat-absorbing surfaces;

— conservation of heat in the heating appliance and minimization of heat dissipation.

The overall efficiency (E) of a cooking process could be specified as follows:[5]

$$E = \frac{\text{Energy transferred usefully to water/food in pot}}{\text{Total energy expended (i.e. the chemical energy content of the fuel)}}$$

This can be seen as having two components, i.e. $E = E_H . E_S$ where:

$$E_H = \frac{\text{Energy transferred usefully to water/food in pot}}{\text{Total energy input to the cooking vessel}}$$

$$E_S = \frac{\text{Total energy input to the cooking vessel}}{\text{Total energy expended}}$$

In other words, $E_H$ could be seen as pot efficiency and $E_S$ as stove efficiency.

Inefficiencies may usually be traced to aspects such as:

— Incomplete combustion of wood due to inadequate or unevenly distributed air;

— heat loss from the stove due to conduction from the stove wall, convection of hot gases, and radiation from the hot surface of the stove;

— inadequate heat transfer from flame to pot due to the surface area, shape, size, colour, etc. of the pot;

— other aspects such as the heat needed to first warm the stove, loss of heat if firewood is left to smoulder after cooking is complete, etc.

The principle underlying the stove improvements is to regulate the inflow and outflow of air currents in such a way as to ensure the

*An improved stove with chimney, pot hole covers,*
*and firebox doors: India*

efficient combustion of wood and make maximum use of the heat
generated. Improvements over the open fire usually include one or
more of the following: the regulation of air flows through a system of
flues* and the prevention/minimization of heat loss by means of a
fire-box door, flue dampers,** covering of pot holes when not in use,
ensuring that the pots fit the holes well, the addition of a chimney,
and so on.

Different experiments give different emphasis to various aspects of
possible improvements. One distinction often made is between stoves
with flues and those without. It is generally claimed that stoves with
flues are more efficient because flues improve combustion by drawing
more air through the wood. However, flues also increase the distance
between the flame and the pot which might lead to a loss of heat. In
practice, tests on the superiority of flues have been known to give
variable results. Some studies (e.g. Singer, 1961) show that flues do
increase stove efficiency, but others (e.g. Acharjee, *et al.*, 1974a,
1974b) find the opposite.

Other aspects of stove improvement emphasized by some

---

*Channels carrying combustible gases; chimneys are vertical flues.
**Means of controlling the air flow within a flue.

experimenters[6] relate to whether or not the flue is tapered at the top—tapering is said to decrease efficiency, and what the slope of the duct is—adding a slope is said to decrease efficiency. In addition, the pot characteristics are important: flat-bottomed pots and pots which are black on the outside are supposed to be less efficient. The material of which pots are made similarly affects overall efficiency (Islam. 1980: 98-99).

Such improvements generally require technical measurements, such as temperature measurements of the stove, the flame, the kitchen air, and the pot surface, and measurement of the calorific value of the fuel and residues (see Dutt, 1978a: 8). If the stove has flues then flue losses are estimated from the flue surface temperature. These measurements would help to provide guidelines on the overall stove design.

However, testing the effect of specific design improvements on stove efficiency, when the stove is used *in practice*, is by no means a straightforward procedure; and the way in which such tests are usually conducted often makes suspect many of the stated claims of improvement.

Consider, for instance, the two standard procedures for testing a stove: heating water to boiling point and cooking a 'typical' meal. The water heating test has the advantage that it is simple to perform and easy to replicate. However, determining the efficiency of wood-burning stoves, even with this simple test, can be complicated. The major problem lies in the difficulty of controlling the heating rate in the stove (which relates to how quickly and how high the flame temperature rises and the efficiency of heat transfer from the fuel to the pot). Technically, the higher the heating rate the less fuel is needed to heat water or to bring it to boil. But because the heating rate cannot be controlled easily (due to the difficulty of ensuring that the wood used is the same in all respects each time the experiment is performed, and that all other conditions are the same as well[7]), even a standard operation such as heating a given quantity of water, in a given vessel, in a given stove, can give variable results.

It might of course be possible to establish, through a large number of tests, which stoves, on average, have a high heating rate and which a low one. But a second problem still remains, namely that a stove which might prove efficient in the water heating test may not prove so in the cooking of a meal, since cooking a meal needs a range of heating rates. For example, in cooking rice a high heating rate would be desirable for boiling the water and a lower one would be adequate to keep it simmering. Hence a stove with a high heating rate which performs well in the water-heating test, may not perform as well in cooking rice. In other words, two stoves, tested in terms of their relative efficiencies in water heating, can reverse their positions when tested in terms of a 'standard' meal. This was, in fact, noted in India, where it was found that the improved South Junagadh stove was more efficient than the one it was being tested against, in terms of heating water, but less so in cooking an identical meal (GOI, 1964). And ultimately stove efficiency needs to be established in terms of actual cooking performance.

Other complications in testing the efficiency by cooking a meal relate to variations in what constitutes a 'meal' and the way it is cooked. The type and variety of *foods* cooked, the sequence in which they are cooked, the degree to which they are cooked (e.g. well-cooked or lightly prepared), the form in which they are cooked (e.g. fried, baked or boiled), and the quantities in which they are cooked (a doubling of food quantity will not double fuel requirements) would all affect the quantum of fuel needed. Additionally, the amount of fuel used can vary a good deal according to the skill of the operator.[8]

In other words, the concept of 'efficiency' as applied to wood-burning stoves is itself complicated and efficiency tests on such stoves are difficult to perform and do not give uniform results.[9]

But practical difficulties apart, the way in which many efficiency tests have actually been conducted has several shortcomings which could have been taken care of. These shortcomings relate to the following aspects:[10]

(i) Control procedures followed: the stoves compared are not always the same in all respects other than the specific modification whose effect is being tested. This would affect the reliability of the result obtained. For instance, it is noted (see Dutt, 1978b) that Singer when testing the effect of flues by comparing two types of Indonesian stoves—one with flues and one without—did not ensure uniformity of all other aspects: the stoves with flues had two or three pot holes, and the ones without flues had one pot hole. Hence his contention that stoves with flues are more efficient cannot be seen to have been established from the test performed.

(ii) Sample size: most studies do not give their sample size, but appear to have quite small samples (Dutt, 1978b). The reliability of the conclusion can depend crucially on this since, as already noted, there can be considerable variability in the results obtained even on the same stove for the same type of test.

(iii) Heating rates: tests are usually done at the highest heating rate for gas, electric and kerosene stoves while, as noted, cooking a meal usually needs several heating rates.

Hence the claims made about increases in efficiency as a result of improvements undertaken, may or may not hold, or may not hold to the extent stated. It is estimated, for example, that simple improvements in stove design could reduce firewood requirements considerably—perhaps by half that needed with the open fire (Evans, 1978, on a Guatemalan stove; and Draper, 1977, on an Indonesian stove); or at least by a quarter of the open fire requirements (Makhijani, 1976: 21, on the smokeless, Indian HERL *chulha*). Expressed differently, relative to the approximately 6 to 10 per cent efficiency of the open fire, cooking efficiency with improved stoves, it is claimed, can rise by 23-29 per cent (Singer, 1961, on the Indonesian stove), or even 38 per cent (Draper, 1977, on the UNDP stove project in Indonesia). Improvements in pot design are supposed to further decrease wood needs: Singer estimates that the reduction could be by 30 per cent relative to the old pots, and claims that the total effect of improved stove design plus pot design has helped reduce daily firewood needs of an Indonesian household by 65 per cent and cut cooking time by 25 per cent. However, in the light of the earlier discussion all such claims about improvements in stove efficiency need to be viewed with caution.

Also, underlying these claims of a reduction in firewood requirements is an assumption of *ceteris paribus*, that is, everything else remaining the same. In reality, many poor households live at such a low level of energy consumption that a more efficient stove that requires less wood for an existing pattern of consumption, may in fact bring about no overall decrease in firewood use: the same amount may continue to be used but the household may now be able to cook more food and/or cook more nutritious food. Of course this too would be highly desirable, as it would constitute an improvement in the welfare of such households.

Further, relative to the open fire, stoves have other potential benefits such as a reduced risk of burns or scalds (from the sparks of burning logs or unstable pots) both for the women cooking and the young children playing around; lesser fire risk in huts made of easily combustible material such as straw, wood, etc.; the elimination of health problems associated with smoke-filled kitchens; the saving of labour otherwise needed to clean soot-blackened pots, walls and clothes, and so on. As will be seen from the discussion in Chaper IV, while many of these advantages have been realized by users of improved stoves, in the sphere of wood saving the results are still ambiguous and unsatisfactory.

## 2. Tree-Planting Schemes

Forest management involves the protection and enrichment of existing forests as well as the undertaking of new tree-planting schemes. Among the underlying principles in protection and enrichment would be the maintenance of a long-term balance of tree cutting and regrowth (sustained-yield forestry), and the recognition of the multiple-use aspect of the forest. New tree-planting schemes may be undertaken to provide a specific set of products (fuelwood, fodder, fibre, fruit, timber, etc.), to reclaim land which has been overgrazed and is subject to soil erosion, to protect the environment in general, for beautification of the landscape, and so on. A scheme could be planned to simultaneously serve more than one of these purposes.[11]

The technical aspect of new tree-planting schemes broadly involves two issues: first, the choice of tree species, and second the way in which they will be planted, that is, on their own, or together with agricultural crops—agri-silviculture, or together with fodder crops or grass—silvipasture. The planting itself could be undertaken on private land by the individual owning/controlling the land, and/or on government or community land by the government or by the village community.

The choice of trees—especially if trees are chosen for their fuel value as firewood or charcoal (and these would not always be the most appropriate for use as timber or to provide other forest products)—would ideally require the species to have as many as possible of the following qualities (see Burley, 1978):

> survival—resistance to adverse environmental factors such as drought, soil salinity, insect attacks, etc.;

fast growth—yielding a high volume of fuelwood per hectare per year and short rotations;

easily available low-cost seed;

strong coppicing ability—this reduces the cost of establishing and tending the second crop;

easy to handle—free from excessive branchiness, thorns, etc. (though in some places thorns may be considered advantageous in protection from cattle);

soil enriching—especially nitrogen-fixing;

yielding multiple products.

In addition, the suitability of a particular species will depend on the climatic and soil conditions in the region where it is to be planted, as well as the specific use for which trees are needed. Attributes such as the rapidity with which the wood burns, and the heat of fire obtained would also be significant considerations for users. In Central America, for example, the oak is noted to be favoured over cedar because 'oak burns hot and clear without smoke . . . cedar gives a hot fire but throws off dangerous sparks' (DEVRES, 1980: 27). Here pine is used for kindling the fire.

In choosing suitable species, however, non-technical considerations would also need to be taken into account. For instance, superstitious beliefs surround certain trees in certain communities, and such trees are thus unlikely to be grown by the villagers: in Senegal, cashew trees are believed to be the abode of ghosts and there are similar associations with the banyan-tree in north India. For this reason, as well as for taking advantage of indigenous knowledge about trees and ensuring that the trees chosen will fulfil local needs, the involvement of the local people in tree selection also becomes important.[12]

Unfortunately, in most reforestation schemes, even the technical aspects relating to the choice of appropriate species are not given the needed attention. The promotion of eucalyptus in many such schemes across the Third World without an adequate prior consideration of its suitability is a case in point. This tree species is fast growing and high yielding. According to Uhart's (1976) assessment, the annual increment expected per hectare ranges from 4-5 m³ for the drier areas to 20 m³ for the wet ones. The figures Earl (1975: 47) provides are higher. He notes that 20-70 m³ per hectare per year would be representative of South America and 15-25 m³ of

Smitu Kothari

*Unsuitable choice: imported Japanese pine planted in the hill region of West Bengal, India, falls after a heavey storm.*

tropical Africa. Under exceptionally good conditions, increments of over 50 m³ are noted to have been obtained in eucalyptus fuelwood plantations in Africa (under 6-8 year rotations) although on drier sites figures might be below 10 m³.

However, fast growth and high yields do not always provide adequate justifications for the choice of a species, and the large-scale promotion of eucalyptus has in fact become a subject of major controversy in recent years in India. Those opposing its spread raise mainly two types of objections:[13] first, it is argued that the species is ecologically destructive and especially inappropriate for dryland ecosystems. It is said to be highly water consuming and to rapidly suck up groundwater,[14] thus reducing water availability for crop production in the nearby fields; it is also noted to inhibit the germination of seeds of many other plant species, to have roots which exude toxic chemicals that adversely affect the soil, and to have leaves that do not decompose easily and therefore, cannot replenish the topsoil. Secondly, the tree does not provide an answer to the fuel and fodder needs of the rural areas: its wood burns much too quickly to be suitable for cooking, and animals do not eat its leaves.[15]

In answer to some of the technical criticism, forestry officials and others argue[16] that eucalyptus allows more rain water to reach the ground and infiltrate the soil relative to several well-known local tree species such as teak and *chirpine*; that undergrowth is prevented even more by other trees such as mango, banyan etc; that the tree acts as a wind-break and so helps increase crop yields; and that there are over 500 varieties of the tree of which those growing in India do not consume much groundwater.[17] Be that as it may, the criticality of these factors would of course vary by ecological zone and context: near canal banks, for instance, eucalyptus could help decrease soil salinity; and in Israel it has been used to reclaim swamps (Douglas and Hart, 1976: 154). What is important is the need to gather detailed technical data regarding the ecological implications of eucalyptus and other tree species in given regions, before they are promoted on a large scale. Also, in the context of the present discussion, the *social* usefulness of the tree would need to be a primary consideration in the choice.

Among trees commonly recommended for fuel plantations are *prosopis* and *leucaena*. Unlike eucalyptus which has a high water requirement, *prosopis* is drought and salinity tolerant, and hence said to be especially suited to semi-arid regions. Additionally, it is fast growing, has thorns which protect it from grazing cattle, and it coppices well. It grows from plant to harvestable size in 5 years and has an average yield of 154 tonnes of wood with 30 per cent moisture per hectare per year (see Bokil and Rao, 1979).

*Leucaena* (also called *subabul* in India and *ipil-ipil* in southeast Asia), a leguminous tree-shrub which fixes nitrogen, is multipurpose in nature and would belong to the category of a tree crop since it provides both fodder as well as fuelwood. The practice of growing trees to provide fodder for livestock, or food for people (fruits, nuts, etc.), in addition to wood, has in fact been and continues to be common in many parts of the world. Douglas and Hart (1976) list a variety of such tree species: for example, *Indian beech, jehab-nut, carob, velvet tamarind, tree-lucerne* (all of which are leguminous) and sal are some of the trees grown in parts of Asia and Africa that provide feed for livestock (and sometimes for humans as well); and *namnam, parkia,* almond, chestnut, cashewnut, *bokhara plum* are some of those that provide nuts and fruits (or pods which can be ground into meal or flour) for human consumption.

Many more such species could be identified and a large volume of

literature on the trees that would be suitable for different uses and to
different ecological conditions already exists.[18] However, the aim
here has not been to provide a check-list of species but to highlight the
type of considerations which would enter into the choice of species
appropriate for a community's needs.

Apart from the choice of trees, the way in which they are
planted—that is, whether on their own or along with food/fodder
crops—can be seen to have both advantages and disadvantages. Earl
(1975: 47) lists some of the advantages of plantations such as:
(a) they allow close control of wood quality and yield a crop with a high
percentage of harvestable woody growth, and (b) yields are much
higher than under natural forest conditions. Since the crop is
homogeneous and fast growing, a high out-turn of harvestable wood
per hectare is possible. This is of particular advantage in land-scarce
countries; it also makes harvesting easier. On the other hand,
plantations being monocultural are more susceptible to insect attacks
and fires than mixed forests, are ecologically often undesirable, and
would not provide the diversity of forest products that mixed forests
provide.

As a variation on the plantation theme, some have also suggested
fuel gardening, that is, intensive cultivation of wood on small plots
(Romm and Seckler, 1979). This, however, usually requires soil that
is well worked, kept weed free and intensively fertilised. The practice
is in many ways similar to growing vegetables in the garden, but with
less water requirements.

Intercropping is often suggested as a technical solution to land
scarcity and to the fact that land has alternative uses for cattle grazing
and agricultural cultivation. The practice has an advantage over
plantations in terms of both the multiple use of land as well as in
providing the means of combating a decline in soil fertility.
Silvipasture—the planting of trees with fodder, or the allowing of
controlled grazing on forest vegetation (grasses, weeds, twigs and
brush, etc.) and agri-silviculture—the growing of trees along with
agricultural crops, are the two common forms of such intercropping.

Agri-silviculture has been noted to be a traditional practice in
several tribal communities in parts of the Third World. For example,
Openshaw and Morris (1976) describe how under the *vihamba*
system practised by the Chagge coffee farmers, in the upper zone of
peasant farming on the slopes of Mount Kilimanjaro, a three-tier
system of farming is practised—trees at the highest level, an

intermediate zone of coffee and bananas, and a groundcover of legumes. Likewise, the *Meru* tribe of Kenya grows trees with bananas, coffee and *khat*. However, much of this agro-forestry tends to be *ad hoc* and does not constitute planned rotations.

More recent and systematic attempts to promote agri-silviculture generally involve the temporary settlement of migrant farmers or shifting cultivators on forest land; they are allowed to farm the land between the trees until such time as the trees shade the land completely. The preference is for quick-growing trees, especially those leguminous in character.

For a variety of socio-economic reasons, however, the schemes have run into difficulties—an aspect which will be spelt out in Chapter V.


### 3. Improved Wood-Conversion Hardware[19]

Wood can be converted into a variety of secondary fuels—by carbonisation (charcoal), distillation (methyl alcohol, etc.) and gasification (producer gas,* water gas** and liquid fuels). Of these, it is really only charcoal which can be considered a close substitute for the wood used in domestic cooking. The other woodfuels are essentially used for industrial purposes, and are outside the scope of the present study.

Charcoal has both advantages and disadvantages relative to wood as a domestic fuel. The main advantages lie in its smokelessness, its having a higher calorific value, and its being easier to store and less bulky to transport than wood. Transportation costs increase with distance and beyond a certain distance charcoal becomes much more economical than fuelwood.[20] The conversion of wood to charcoal is also seen to have the advantage of enabling the utilization of wood-waste generated in sawmills or of the off cuts left when trees are felled for timber.

The main disadvantages of charcoal lie in its fragility and hence the ease with which it is broken, and in the danger of carbon monoxide poisoning in the absence of free air circulation. Also, there is a loss of total usable energy in the carbonisation process: for example, taking

*A mixture of carbon monoxide and nitrogen
**A mixture of carbon monoxide and hydrogen

one tonne of air-dry wood as containing 3.5 m kcal, and one tonne of charcoal as containing 7.1 m kcal and assuming a 30 per cent yield of charcoal from wood, there would be a net loss of 1.4 m kcal of usable energy per tonne · of firewood converted into charcoal. The advantages of charcoal, as noted, lie elsewhere, especially in the easier transportation possible over long distances (see Earl, 1975: 27-31).

A variety of hardware can be used to convert wood into charcoal: kilns, continuous kilns, furnaces, retorts – the general principle being 'combustion of part of a pile of wood until it is hot enough to be able to react exothermically in a limited air supply, i.e. to carbonise' (Earl, 1975: 29).

Technically, the simplest device is the earth kiln, consisting either of a hole in the ground (covered with turf) in which the wood is fired (the pit kiln), or of a stack of burning wood (at ground level) covered with turf or earth. It requires no financial expenditure on equipment. The charcoal output is estimated to be about 8 to 12 per cent of the weight of dry wood by the earth-stack method and 12 to 15 per cent by the pit method (Uhart, 1976: 4, 5). However, relatively large pieces

*An earth kiln: cheap but wasteful*

of wood are needed, the charcoal produced is not of high quality, yields are low, and the time taken to carbonise is long and difficult to control.

Improved kilns are made from brick, concrete, or even metal—as, for example, are portable steel kilns. With steel or masonry kilns it is claimed that charcoal yields can be raised to 40 to 50 per cent.[21] Portable kilns have the added advantage of mobility over fixed-location kilns, being especially useful where the wood available for charcoal manufacture is dispersed widely within a forest. Owing to their size and cylindrical shape it is easy to roll them from one area to another, thus minimizing the distance over which wood would need to be carried. In this way they can also help to utilize timber which might normally be left to rot on the forest floor. Further, such kilns have a short turn-around time and are easy to operate.[22] However, their high cost (about $1000 per kiln[23]) would tend to limit their widespread use.

Continuous kilns usually consist of a vertical, insulated steel cylinder in which the raw material is put in at the top and charcoal

D.E. Earl/FAO

*A portable charcoal kiln: efficient but expensive*

withdrawn from the bottom. While such kilns provide a continuous supply of high quality charcoal and can use small-sized residues, the process also needs a continuous supply of small wood, constant attention by an operator, and usually an external fuel source to begin carbonisation.

Furnaces are more capital intensive than kilns, and need chipped, relatively dry wood (in kilns generally the moisture content is not critical). However, they are an improvement over kilns[24] in that they can utilize any kind of organic material, including sawdust and bark, and produce a continuous supply of high quality charcoal. Also, although for initiating the process and for driving the charge through, an external energy source is needed, the energy derived from the exothermic reaction of part of the charge provides all the heat needed for continuous carbonisation.

Neither furnaces nor kilns, however, help to capture the by-products (the liquids and gases) of the carbonisation process. For this, retorts which are also highly capital intensive are needed. These are containers to which heat is applied externally until charcoal is produced and the gases generated can be collected, condensed and fractionated—the wood distillation process—to obtain methyl alcohol, acetic acid and pitch. With retorts, the gases generated from the controlled heating of wood and charcoal (namely, producer gas and water gas) can also be collected and used directly as fuel: for example, producer gas is used in internal combustion engines.

From the above it is clear that a variety of technical options exist for obtaining secondary fuels from wood. The suitability of the methods would depend on how high a quality of charcoal is needed, whether the by-products (gases etc.) have a direct use and are worthwhile recovering and, most important, the scale and location of operation (for example, whether produced in one spot or in several places), as well as the form of ownership. In the present context, our interest is essentially in improved equipment (especially improved kilns) for producing charcoal for domestic use, rather than in equipment for capturing other by-products (which as noted have mainly industrial uses). In technical terms the efficiency of kilns can be increased considerably by improving over the earth-kiln method; but such improvements also involve substantial expense and, for the more sophisticated equipment, usually require skilled operators.

In the following chapters and appendix the range and complexity of problems encountered in attempts to promote improved wood-burning stoves, tree-planting schemes and improved charcoal kilns are spelt out.

## Notes

1. Quoted in Foley and Moss (1983: 75).
2. For details see Raju (1953); Prasad (1982); Foley and Moss (1983); Joseph (1980); and Sarin (1983).
3. Designed at the Hyderabad Engineering Research Laboratories (HERL).
4. See for example, the Ministry of Fuel and Power, U.K. (1944).
5. The definitions, slightly modified in language, have been taken from Dutt (1978a: 6-7). The discussion in this section has also drawn upon Dutt (1978b).
6. See Government of India (GOI, c. 1964: 3).
7. See for example Joseph and Shanahan (1980: 3).
8. Sometimes even the attention given to the person cooking during a test can help reduce firewood use. As Wood (1982: quoted in Foley and Moss, 1983: 124) notes: 'In Upper Volta, two weeks of daily measurements of wood consumption in households *without* stoves was alone sufficient to reduce consumption by 25%.
9. Jan Bialy, who in the course of his research on energy use in Sri Lanka carried out some 70 tests to determine how much wood was needed to boil water and cook rice, and whether the fuel needs varied by the pot design and hearth design, came to similar conclusions regarding stove efficiency tests (see Bialy 1979a and 1979b).
10. Also see Dutt (1978b) who has reviewed several studies.
11. Also see Adeyoju (1976).
12. This aspect is discussed in more detail in Chapter V.
13. For instance, see Bandhopadhyay and Shiva (1984); Shiva and Bandhopadhyay (1984); Bahuguna (n.d.); also see Agarwal (1983b) for a brief overview of the controversy.
14. By one estimate a single eucalyptus tree can use upto 80 gallons a day and is thus unsuitable for areas with limited amounts of groundwater (Douglas and Hart, 1976: 154). Bahuguna (n.d.) describes several cases in the Indian context of water shortages created by eucalyptus that he has observed, such as the drying up of handpumps in a village in Himachal Pradesh, and of swamps in Uttar Pradesh where wild animals came for a drink.
15. This often makes it a favourite species with foresters since it needs no special protection from animals.
16. For instance, see Gujarat Forest Department (1984), and Jha (1983a).
17. On this point Bandhopadhyay and Shiva (1984) argue that in practice plantations in India are based exclusively on eucalyptus hybrid which can lead to destabilisation of vulnerable dryland ecosystems.
18. Huria and Achaya (1983) for instance provide a detailed listing of tree species that would provide fuel/timber/fodder/food, etc. and be suitable for specified ecological zones in India. Agarwala, *et al.*, (1982) also discuss the suitability of selected species for different regions of India. The National Academy of Sciences (1980), again, lists and describes a large number of firewood-yielding shrubs and

tree species that could be grown in different parts of the Third World.

19. This section draws heavily on Earl (1975) and Uhart (1976).

20. Earl (1975: 74) compares the cost of producing and transporting fuelwood with that for charcoal in the context of East Africa and comes up with an indifference point of 82 km. He notes: '...at distances less than 82 kilometers the marginal value of heat from fuelwood is higher than that obtainable from charcoal and at distances greater than 82 kilometers the position is reversed.'

21. See Hughart (1979: 50).

22. See Paddon and Harker, (1979: 3) on a number of these points.

23. See Hughart (1979: 50).

24. Earl (1975: 29) mentions the Herreschoff furnace in particular.

# CHAPTER III

# THE DIFFUSION OF RURAL INNOVATIONS: SOME ANALYTICAL ISSUES

*Change agents should be like waves on a sea; made of the same water, but which rise up above the water according to the needs of the situation and merge into the water again when the need is over.*

—a Filipino peasant[1]

*We are like fish in a polluted sea. . . . I find that no matter how hard I try, I cannot swim as fast or as properly as I should. . . . This happens not because something is wrong with me, the fish; I know rather, that something is wrong with the sea I swim in.*

—a Filipino peasant[2]

From the theoretical and empirical literature on the diffusion of various innovations in the rural areas of Third World countries, a spectrum of approaches (often varying by academic disciplines) to the diffusion process can be gleaned. In the present chapter a critical review of the differences in these approaches is undertaken and it is argued that not all approaches would be suitable equally for the diffusion of every innovation. Innovations differ from one another in what could be termed their technical, economic and social characteristics, and the likely effectiveness of a particular approach to the diffusion of a particular innovation would depend on these characteristics. A typology of innovations based on these characteristics has been attempted here. This provides the analytical framework for subsequently considering the case of woodfuel innovations.

## 1. Approaches to the Diffusion of Rural Innovations

A review of the literature on the diffusion of rural innovations in Third World countries reveals a wide range of studies covering a variety of innovations: new agricultural practices, High Yielding Variety (HYV) cereals, contraceptive technology, health technology, and so on. Their essential differences, however, may be seen to lie in what are considered to be the main factors constituting bottlenecks to or catalysts in the diffusion process.

Broadly, these factors are seen to concern the following (often interrelated) aspects: the attitudes and personality traits of the individual adopter; the physical attributes of the innovation and the method of its first generation and subsequent development; the economic costs/benefits associated with it; the supporting rural infrastructure; and finally the socio-economic structure of the community.

In numerical terms, probably the largest number of studies would belong to the approach best characterized by the work of Rogers and his 'school'.[3] This approach takes both the need for the innovation and its attributes as given, and concerns itself primarily with the process of communicating information on the pre-developed innovation to the final users. Insofar as the potential users recognize the need for the innovation, the process of diffusion reduces to arranging an information delivery system (mass media, extension agents, demonstration trials and so on).[4] However, for those who do not recognize the need (those lacking in 'venturesomeness', the 'sceptics', etc.), the diffusion process would also require overcoming their 'scepticism' and 'persuading' them to change their 'attitudes'. Here the role of the informal, interpersonal communication channels or 'network' is taken to be primary, as is the role of the 'change agents' and 'opinion leaders' within it.

The extent and pace of diffusion (and hence its success or failure) are thus seen to depend on the one hand on the personality characteristics of the potential adopters, and on the other hand, on the efficiency with which the 'network' channels can function. This functioning, in turn, is seen to vary with the degree of 'traditionality' or 'modernity' of the 'social system', the latter being defined as 'a collectivity of units which are functionally differentiated and engaged in *joint* problem solving with respect to a '*common* goal' (Rogers and Shoemaker, 1971: 28; emphasis mine). Rogers, while allowing for the possibility of hierarchies within this social system, sees them essentially as influencing the individual's behaviour—his/her response to 'communication stimuli'—and not the individual's ability to adopt. Interpersonal relationships within the social system are thus seen as being complementary rather than antagonistic in nature.[5]

In short, by this approach (which could be termed the 'straight transfer' approach), problems of diffusion are basically seen as problems of information-communication and persuasion. Aspects such as the unsuitability of the innovation itself, or difficulties arising from the material conditions (rather than the personality traits) of the potential adopter, are little emphasized. Also, the relationship that is implicit in the approach, between those seeking to promote the innovation and the potential adopters, is unequal and hierarchical: the promoters are seen as the ones with superior knowledge and the rural poor as those who do not know what is good for them.

In contrast, a number of other studies recognize that the user's decisions are usually based on rationality, subject to social, economic and cultural specificities. These studies place primary emphasis on the process by which the technique itself is generated and developed, and on the need to ensure its suitability to the user's requirements. They point out that the distinction between innovation generation and diffusion is a false one, and that the innovation cannot be taken as exogenously given but must be developed/adapted in the field itself. The feature common to most such studies is their emphasis on the desirability of close interaction with, and involvement of, the final user in the innovation process itself. Where they differ is in the degree of involvement envisaged.

In some, the need for user-involvement is admitted essentially in the final stages of the innovation process, that is, a prototype of the innovation would have been developed in the laboratories/research stations and then adapted to the environment (especially physical) of the users. Griliches' study (1957, 1960) on the diffusion of hybrid corn, although relating to the USA, is of interest here, since it presents one of the earliest attempts to incorporate adaptation to varying ecological conditions as an essential component of the diffusion process. A variant of this approach is implicit too in Rosenberg's (1975: 29) following statements: 'Innovation is simply the beginning of the diffusion process' and 'the diffusion process is typically dependent upon a stream of improvements in (the) performance characteristics of an innovation, its progressive modification and adaptation to suit the specialized requirements of various sub-markets'.[6]

Others emphasize that a prototype of the innovation can be obtained from the users themselves and then given sophistication by the scientists in the laboratory. They argue that users often generate innovations, or undertake innovative adaptations, which might lack technical sophistication but which are significant in that they directly manifest users' needs, and embody a store of indigenous knowledge and skills which should be brought into use. This assumption underlay the attempt in Meiji Japan to involve the farmers in the innovation process: 'Intimate knowledge of the best of traditional farming methods was thus the starting point for agricultural research and extension activities' (Johnson, 1969: 61[7]). This again was the idea underlying the promotion in China of the 'three-in-one' innovation teams (a combination of workers, technicians and management personnel) within factories during the Cultural Revolution (Dean, 1972: 531), and of a close relationship between peasants and Research and Development (R & D) personnel during the early 1970's (Ishikawa, 1975). These attempts are in sharp contrast to the one-way flow of information under the 'straight transfer' approach.

Adaptation of innovations, whether in consultation with users, or with the help of users, or taking user innovations and then adapting them, is also seen to have additional advantages, viz:

(i) Preventing indigenous skills and knowledge from dying out: this could happen when outside knowledge legitimized by the superior status of the 'experts' (scientists, planners, extension agents) undermines and destroys the confidence and ability of the local experimenter/innovator (Howes and Chambers, 1979: 7), or if indigenous skills fade away through lack of use (Bell, 1979: 47).

(ii) Helping to further develop indigenous skills and knowledge: this, in turn, can enhance the future possibilities of indigenously-generated innovations through the 'learning-by-doing' effect (Cooper, 1979a: 404), or through the release of the 'latent, creative and managerial energy of the farmers' (Hapgood, 1968: 10). Additionally, it can enable the users to gain a better technical understanding of the innovations initially generated outside, and this would give them a greater control over, and involvement in, the process which changes the technical basis of their lives (Bell, 1979: 47, Herrera, 1975: 44).

(iii) Ensuring that the innovation is appropriate to users' needs: this together with the users' increased sense of involvement and understanding of the technical aspects underlying the innovation could bring about a more ready acceptance and hence successful diffusion of the innovation.

However, in this context, some studies while emphasizing the need for innovation adaptation, point out that the ability of a country to successfully undertake adaptation would depend on the working of its formalized R & D system. Hayami and Ruttan (1971: see especially 197-198, 212-214), for example, emphasize that a significant factor affecting the diffusion of HYV cereals in many parts of Asia was the degree of responsiveness of local research agencies to changing factor prices (in this case fertiliser/grain prices), and the ability of the agencies to adopt crop varieties so as to use more of the abundant (cheaper) factor and less of the scarce (expensive) factor. They attribute the 1960s lag in HYV adoption in Asia to an inadequate development of local R & D facilities.

Other studies, however, which also deal with the diffusion of HYVs give greater importance to the user's ability to gain access to complementary inputs (fertilisers, irrigation), and to credit. They argue, essentially on the basis of the vast volume of literature relating to attempts to popularize the improved crop production technology, that institutions providing information, credit, production inputs, etc. are so dominated by the interests of the few who are economically and socially powerful, as to preclude the majority of the people from access to the innovation.[8] In other words, even if the innovation is technically suited to a user's needs, the user may still not be able to adopt it if s/he belongs to an underprivileged section of society.

The degree of institutional transformation seen as necessary for overcoming these biases again varies between studies. Some merely point to the inadequacies of the extension system, the attitudes of the extension workers (Leonard, 1977) or the constraints (diversity of functions, inadequate training, frequent transfers, low pay scales) that define the work conditions of these people (Lele, 1975; IADP, 1966). Others emphasize the rigidity of bureaucratic rules and procedures, the red-tape, and the hierarchical structures of bureaucracies (IADP, 1969; Moulik, 1979; Montgomery, 1965). Only a few point to the class basis of society (which might govern both attitudes and institutional working), and question the feasibility (if not relevance) of bringing about piecemeal changes in the working of specific parts of the State apparatus (e.g. Dasgupta, 1977, Hapgood, 1965).

To sum up, we note from the above that studies dealing with the diffusion of rural innovations vary widely in what they emphasize as being the main hindrances to or aids in the diffusion process. This variation in emphasis may be seen on the one hand to result from differences in individual judgement, and on the other hand, to relate to the type of technology being promoted. Reserving judgement for the moment, a classification of rural innovations by some analytical categories would be helpful in gaining a better understanding of the likely importance of different factors in the diffusion of particular innovations. In the section below such a classification is attempted.

## 2. Some Analytical Distinctions between Rural Innovations

The suitability of a particular approach in the diffusion of a particular innovation may be seen to depend on what could be termed the technical, the economic and the social characteristics of the innovation.

For example, the technical (physical) characteristics become important in determining the extent to which a technology can be generated or adapted in the field rather than in the laboratory (depending, for example, on the material components needed for its development), and by the users themselves rather than by the scientists (depending, for example, on the users' familiarity with the technology or process which forms the basis of the innovation). Thus the possibility of generating contraceptive technology or vaccines or designing watches, radios, etc. outside the laboratory, clearly would be limited, while that of field adaptation of crop varieties, tree species and methods of performing agricultural tasks would be considerable.

Again, the kinds of problems likely to be faced with diffusion and the appropriateness of any specific approach, would be related to the economic and the social characteristics of the innovation. One way of defining the economic characteristics would be in terms of (a) the form—financial or non-financial—in which the costs are incurred and benefits received by the adopter*; (b) the level of these costs and benefits; and (c) the speed with which the benefits can be realized. One way of defining the social characteristics would be in terms of who the potential adopter is—whether it is the individual or the community; and if it is the individual, then of what class and social (e.g. gender, caste) grouping.

The economic and social characteristics together provide a possible way of classifying rural innovations. For purposes of illustration consider the following examples:[9]

(i) those representing a private financial cost and yielding mainly a private, financial, production benefit**: such as HYVs, mechanical agricultural equipment (tractors, threshers, private tubewells), etc.;

(ii) those representing a private financial cost and providing mainly a private, non-financial, consumption benefit: such as watches, radios, etc.;

(iii) those representing a private financial and/or non-financial cost and providing mainly a private, financial, savings benefit: such as family-sized, private biogas plants, which would help save on purchased fuel such as kerosene, and on fertilisers through the slurry produced;

(iv) those representing a social/communal cost—financial and/or non-financial—and providing mainly a financial, production benefit to the individual: such as irrigation canals, irrigation water reservoirs, etc.;

(v) those representing a social/communal cost—financial and/or non-financial—and providing mainly a non-financial, consumption benefit to the individual: such as piped drinking water, public medical care, etc.;

(vi) those representing a social/communal cost—financial and/or non-financial—and providing mainly a financial and/or

---

*Non-financial costs could be in the form of labour time put in for a task, say for building a stove; non-financial benefits similarly could relate to the 'prestige' of owning an article, or an increase in leisure time, etc. In other words, they would cover all costs and benefits where no monetary transactions are involved.

**Private costs/benefits: those relating to the individual (person or household); social costs/benefits: those relating jointly to a community of people.

non-financial savings benefit to the individual: such as contraceptives which save on the cost of rearing children both in financial terms (food, education, etc.) and in non-financial terms (work time); and environment conservation projects (social conservation, flood control, reforestation, etc.).

These represent some broad illustrative categories. Basically, the list could be extended quite easily to cover other, and more complex, combinations of the economic and social characteristics. In practice, some innovations would fit into more than one category, depending on circumstances. For instance, clean piped drinking water in so far as it substitutes for polluted river water, or is introduced in water scarce areas, provides essentially a non-financial consumption benefit (as indicated above); but if alternative sources of water are distantly located, it also provides a non-financial savings benefit (the saving of water-fetching time). Further, to the extent that it improves the health and hence the productive capacity of the individual, it could provide an indirect production benefit in the long run. Similarly, whether the costs/benefits of an innovation are mainly financial or non-financial in nature could vary by context. The essential point is that the nature of the characteristics is likely to define the ease or difficulty of diffusion, and an identification of these characteristics within a given context would provide clues on the appropriate approach for diffusion in that context.

To elaborate, the effects of these characteristics on innovation diffusion would be manifest through: (a) the potential adopter's perception of the advantages of the innovation, and (b) the potential adopter's ability to gain access to its benefits. Consider first the issue of perception. In terms of the economic characteristics it can be suggested that the advantages of innovations such as HYVs and irrigation, which provide a direct, high, financial benefit, and in a relatively short time, are likely to be perceived more readily than those of innovations such as contraceptives or conservation projects whose benefits to the individual would generally be indirect, non-financial and often (for conservation projects) realizable only after a considerable period of time. Again, in terms of the social characteristics of an innovation it can be suggested that the advantages of innovations which men use are likely to be more readily perceived than those used by women, where decisions on cash expenditure are made by men.[10]

*FAO*

— *Women contribute substantially to field labour but do they get access to the cash income generated?*

Next consider the problem of the potential adopter's ability to gain access to the benefits of the innovations. This again relates both to the economic and the social characteristics of the innovation. For example, innovations which require cash expenditure (the economic characteristic) are likely to be much more difficult to diffuse, particularly where the potential adopters are poor and have little cash at their disposal. Again innovations which require communal cooperation for successful adoption (the social characteristic) are likely to be much more problematic to promote. Here the basic difficulty is that the way individuals act within a group would depend on what assumption each makes as regarding how the others will act, and on the degree of assurance that the burdens and benefits of the effort will be shared equally. In the absence of such an assurance, cooperation would usually be difficult to bring about.[11]

These analytical distinctions between innovations give clues about the likely importance of different factors in the diffusion process. For example, identifying the technical characteristics of the innovation would help determine the technical feasibility (if not relevance) of innovation development/adaptation in the field by the user. An identification of the economic and social characteristics would give an idea of whether the innovation can be promoted on a commercial basis like other items which are financially profitable or which provide direct consumption benefits to the individual adopter, or whether it requires a different approach dictated by the communal nature of the innovation, or by its being aimed at a mass of economically and socially underprivileged individuals or households.

A classification of innovations by these characteristics leads further to a number of practical questions as, for instance: if the technical characteristics are such as to permit, and in fact make it desirable, that the users be involved in innovation generation, then how can such an involvement be brought about? Or, where the potential adopter does not have the cash for the innovation which needs financial expenditure, then how can s/he be enabled to acquire the innovation? Or, where the decision to adopt rests with someone who is not the potential user and who therefore does not perceive the need for it, how can this perception be altered? Or, where decisions to adopt need to be communal in nature, how can the consensus of the community members be ensured?

In fact all these questions, in one way or another, are linked to the issue of social structure. We have noted, for example, that the diffusion of innovations which need communal cooperation, requires an assurance that there will be a fair sharing of the costs and benefits among the community members. However, one can legitimately question whether such an assurance can exist in societies where the existing distribution of material resources and political power is highly unequal, and where the distribution of costs and benefits from new schemes may likewise be unequal. Most experiences of the failure of communal projects which involve heterogeneous (in economic and social terms) groups of people, indicate that the reluctance of the underprivileged to participate in the programme is not located in 'irrationality' but in their specific material and social circumstances.

In this sense, institutional and structural changes may become important even when the innovation is not communal in nature, but is aimed at a large number of individuals rather than a few. For, the ability of many potential users to adopt the innovation, again would be dependent on the extent to which infrastructural facilities (such as extension, credit, etc.), serve *their* needs rather than being biased in favour of a few.

Similarly, when we consider the issue of benefit-perception by the individual household, as noted, divergent gender interests within the household may serve as a barrier to adoption. In this case, the social 'institution' which would need adaptation is the family which 'reproduces' specific attitudes of men and women towards one another.

It could likewise be argued that user-involvement too may need structural/institutional changes, since the success of this approach to diffusion is dependent on the degree to which a dialogue is possible between the scientists/professionals and poor peasants, or between male extension workers and female innovation-users, and dialogue usually requires equality (in attitudes, material conditions, etc.) between those conversing.

In short, the existence of divergent, often antagonistic, interests defined by economic (class) and social (gender, caste etc.) hierarchies is likely to constitute circumstances where the diffusion process cannot be treated merely as being a question of a 'straight transfer', or even solely as one of innovation adaptation, but as one where structural transformation (redistribution of material wealth, change in attitudes) might need to be a significant component of the diffusion process.

Some of the problems likely to be faced with the diffusion of woodfuel technology relate precisely to these issues. Consider the instance of improved wood-burning stoves. At first glance, one might be tempted to put them in the same category as watches and radios, since they usually have a private financial cost (albeit a small one), and provide consumer benefits (for example, smokelessness). Indeed in many attempts to diffuse them, stoves are treated as being no different from watches and radios. Such a categorisation, however, fails to take account of certain complexities specific to wood-burning stoves, which would make their advantages less apparent to the potential adopter. For example, the benefits would usually be non-financial and indirect, and accrue mainly to the household women, whose wood-gathering burden would be reduced and cooking made easier and more pleasant with an improved stove design. However, it is usually men who make decisions on household expenditure,[12] and while men would readily perceive the consumer benefits of items such as watches and radios of which they are the primary users, they are less likely to see the benefits of improved wood-burning stoves. Also, stoves, unlike watches and radios, are less 'visible' and hence would not be seen (especially by the men) as conferring the same social prestige benefits to the household. For these reasons it would be less easy to sell wood-burning stoves in the same way as many other consumer items.[13]

In general, in applying the issues discussed to the requirements of woodfuel innovations, *a priori* we could say that the technical characteristics of these innovations do not necessitate a straight transfer approach and would favour a user-involvement approach. Further, both their economic and social characteristics are such that aspects of social structure are likely to impinge on the possibilities of successful diffusion.

However, any definitive pointers or conclusions can only be based on a consideration of actual experiences. In the following two chapters and the appendix the specific experiences with each of the woodfuel innovations—improved wood-stoves, tree-planting schemes and improved charcoal kilns will be considered.

## Notes

1. Taken from Bhasin (1976: 20).
2. Taken from Avila (1976) who spent ten years working with Filipino peasants and compiled some of the things they said in a booklet called *Peasant Theology*.

3. See especially Rogers (1961, 1977), and Rogers and Shoemaker (1971).

4. A good deal of literature on rural diffusion in fact is concerned mainly with the process of *information* diffusion. Among studies relating to India, see Sen (1969); Gaikwad, *et al.*, (1972); and Kivlin , *et al.* (1968).

5. It is only very recently that some of the earlier adherents to this approach, including Rogers himself, have begun to examine these underlying assumptions and have been seeking to 'modify' the classical model of diffusion' (see Rogers, 1980: especially 6-12).

6. Both Rosenberg and Griliches see innovation adaptation to the user's environment as helping diffusion essentially by increasing the economic profitability of the innovation, which in turn is seen as one of the principal factors affecting diffusion.

7. Also see Hayami and Ruttan (1971: 157), who further note that in Meiji Japan farmers' innovations were often tested and refined in Experimental Research Stations.

8. See Dasgupta (1977) and Byres (1972) on India; Griffin (1971) more generally on the Asian experience; and Hapgood (1968) on experiences in Africa.

9. These cover many of the familiar rural innovations introduced in recent years in Third World countries.

10. The individual's failure to perceive the benefits of an innovation could of course be the result of a whole range of additional factors including misinformation given by the extension agent. However, the concern here is solely with the implications of the economic and social characteristics.

11. Some useful analytical insights into this are provided in economic analysis under game theory by contrasting the outcomes of games such as the 'prisoner's dilemma' with others such as the 'assurance game'. In the former, each individual in the group acts in her/his own interest on the basis of individual rational calculation on the presumption that the others will too, even though a better outcome for all would be attained through joint action. In the assurance game, each person operates on the principle of 'reciprocity' on the assurance that the others will too. Cultural/ideological orientation, or growth in group political consciousness would be factors promoting this reciprocity (see Sen, 1973: 96-98; and McNicoll, 1975: 4).

12. See Chakravarty and Tiwari (c. 1977: 46), Bukh (1979: 29), and Arens and Van Beurden (1977: 45). In both Bukh's study in Ghana and Arens and Van Beurden's in Bangladesh, this was noted to be so even where women made a *financial* contribution to household income.

13. In fact, in many respects, the factors affecting the diffusion of improved wood-burning stoves are likely to be similar to those affecting the diffusion of technologies, such as contraceptives, included in category 'vi' of the typology. In both cases (a) the actual financial cost to the individual adopter is relatively slight; (b) the private cost of not adopting is often perceived to be small by the individual—the saving of firewood gathering time or firewood purchase cost in the one case, and the saving of child rearing time and of child rearing cost in the other; and (c) the status of women is important: women usually bear the burden of firewood gathering, smoky huts as well as of child rearing while men usually decide whether the new stove should be bought (where purchase is involved) or birth control be practised. However, unlike contraceptives, which by their

technical characteristics are essentially laboratory-based innovations, wood-burning stoves offer considerable possibilities for field adaptation and user-participation in the adaptation process. In this sense, they would be closer to innovations of category (i), such as HYV crops, than to those of either category (ii) or (vi).

# CHAPTER IV

# FACTORS AFFECTING THE DIFFUSION OF IMPROVED WOOD-BURNING STOVES

*If they had tried cooking in 1½ ft tall kettles, constantly stirring mush for ten people with a large wooden paddle, they would not suggest waist-high wood stoves (unless they also added step-stools), flat bottomed pans (which burn around the edges) and lids (for pots requiring constant stirring).*

—Hoskins (1979b : 37)

Improved wood-burning stoves (henceforth called 'wood-stoves') through their potential for saving on the amount of firewood needed by the household for cooking, are seen as one way of alleviating the problem of firewood shortages. They can also provide cleaner and more pleasant cooking conditions in the kitchen. Clearly, however, the degree to which this is achieved would depend on the extent to which the stoves do in fact save wood in practice (and not merely in laboratory tests), and the degree to which they are accepted by the mass of rural users, especially by the women of poor households who (as noted) experience the problem most acutely.

Over the past three to four years, more than a hundred programmes for diffusing improved wood-stoves of a variety of designs have been initiated in different parts of the Third World. Only a dozen or so of the programmes currently being implemented are noted to have been running for over two and a half years, and even fewer are believed to have distributed or sold more than 5000 wood-stoves (Joseph, 1983: 2). Partly, but not entirely, because of the newness of these programmes, in-depth evaluation studies on them are extremely few. However, of these, some are sufficiently detailed to provide rich illustrative material and useful pointers. The discussion below draws upon these, as well as upon relevant literature relating to the diffusion experience of other rural innovations, to illustrate particular points.

Applying the analytical framework outlined in the previous chapter to this context, we can say *a priori* that the characteristics of this innovation are such as to make its diffusion a complex process. For example, its primary use is in cooking, an activity whose requirements vary by cultural norms and the specific needs of users; this would make it unsuited to mass production in the laboratory, and would require field adaptation to users' needs. Again the non-financial, indirect form of its benefits—essentially a reduction in women's work burden and labour time (which may have little or no opportunity cost in monetary terms[1]), in the collection of what is customarily considered a 'free' item (firewood), would make it a low priority item, especially where the decision on adoption rests with men. How these and other economic and social characteristics of wood-stoves tend to affect their diffusion in practice is the focus of the discussion that follows.

In broad terms, we could divide the factors likely to affect the diffusion of improved wood-stoves into five (often interrelated) categories: (1) the method of wood-stove designing and

*Need for appropriate designs: in some communities women prefer outdoor cooking—Niger*

development; (2) economic aspects; (3) infrastructural aspects (extension, credit, etc.); (4) cultural aspects (attitudes to change, etc.); and finally what could be seen as the connecting link, viz. (5) social structure. Let us consider each in turn.

## 1. The Method of Wood-Stove Designing and Development

Available evidence strongly points to the unsuitability of a 'top-down' and 'straight transfer' approach for the diffusion of wood-stoves, and to the importance of field adaptation involving the local users, local materials and local artisans.[2] Consider some case studies. Prominent among these is Shaller's (1979) research on the diffusion of the Lorena stove in the severely deforested highlands of Guatemala. This study carried out over a period of three months is based on 36 in-depth interviews with stove owners, plus an intensive observation of the cooking practices of six families in the area where the author and his wife lived. The stove was developed in 1976 at the Estacion Experimental Choqui (ICADA), a small appropriate technology centre near Quezaltenango, and is meant to replace the open fire. Formed from a monolithic block of sand and clay (locally available materials), it is designed to conserve firewood (the promoters estimate a saving of 50%) and decrease smoke build-up in the kitchen. The stove was first developed and then disseminated to the users by private and government organisations.

Shaller does not indicate what the level of initial adoption has been, that is, what proportion of those exposed to the stove have in fact adopted it. He notes an overall high level of acceptance of the stove in that most of those who have adopted it are using it daily, in lieu of the open fire. However, the effectiveness of the diffusion programme *vis-a-vis* one of its primary purposes, namely saving firewood, has been limited. A significant reason for this is that the users, while perceiving that the stove has a number of advantages (indicated below), also see in it several disadvantages which they have sought to overcome by 'adapting' the stove to their particular needs, thereby reducing its efficiency in terms of the wood-saving potential inherent in the original design.

As perceived by most users, the main advantages of the stove over the open fire are the following: decreased smoke (although some households see this as a negative feature, as for instance where the house-roofs are of straw which the smoke-soot helps to seal and make water-tight, or where the smoke serves to get rid of mosquitoes or of pests in ears of corn hung from the rafters); cleaner and more comfortable working conditions (cooking can be done standing up); less effort needed in cooking (with the open fire, the pots are often in a precarious position and need constant watching); and some saving of firewood (although no precise estimates of this were made, and only two families claimed saving half of the wood used previously, while four reported using the same amount as before).

The main disadvantages perceived by the users are that the stove provides no space heating; the cooking surface is inflexible in that the pot holes provided in the body of the stove limit the number and size of pots which can be used; the pots often do not fit the holes (causing smoke and heat to escape), and the stove needs more careful maintenance.

Attempts by the users to 'adapt' the stoves to suit their needs include removing the firebox door to provide some space heating, making almost exclusive use of the firebox to cook individual foods quickly, rather than using all the pot holes for slow simultaneous cooking of different foods as had been intended in the stove design, the removal of the flue dampers due to an inadequate understanding of their functions in controlling and directing heat flows inside the stove, the use of the firebox as an oven (a use for which it had not been intended), and so on. In other words, the women have attempted to keep to the cooking techniques that they had been using with the open fire. Not all the adaptations have been in the nature of adaptations-in-use. In some, the owners have attempted ingenious improvements of the stove design, but adaptations-in-use have been more common.

The need for the users to make the noted 'adaptations', point to specific lacunae in the programme (as apparent at the time of Shaller's study). First, the specificity of users' needs has not been taken into account adequately in the design. If the local methods of cooking had been better understood, appropriate modifications could have been made without a loss of technical efficiency.[3] Second, the stove has been promoted as a piece of equipment rather than as a new process of cooking. Yet subsumed in the design is a somewhat different cooking process than that possible with the open fire. To enable the user to successfully adapt to the design, however, requires making the user more familiar with the basic principles underlying the improvements. The two-day stove building courses that have been held, have not gone into the simple theory underlying the improvements and the

Lorena stove cooking process. Additionally, the participants in the course have been mainly men, while women are the primary users of the stove. Hence women are forced to learn how the stove functions on their own, or get second-hand (and usually inadequate) information from their husbands. That some users have an innate ability to adapt and manipulate the technology for use is apparent from the design modifications they have made anyway.

In other words, if there had been a deliberate attempt to integrate the innovation and diffusion processes, and to closely involve the local users designers in the programme from the beginning, it would have been possible for information to flow in more than one direction, viz.:

— from the user to the outside designer in terms of the user's needs, thus enabling the development of a design more appropriate and satisfactory both in terms of efficiency and user-specificity;
— from the indigenous to the outside designer, thus making an appropriate use of indigenous technical knowledge and skills;
— from the outside designer to the user in terms of basic principles, thus helping to further develop indigenous technical knowledge and skills. In particular, this would have needed the involvement of the village women. As the project had in fact been implemented up to the time of Shaller's study, this had not taken place.[4]

Now consider another example. This relates to an attempt made in rural Ghana to replace the traditional fire by wood-stoves (see Hoskins, 1979b). The model (recommended by the Canadian Hunger Foundation and the Brace Research Institute, and introduced by the Department of Social Welfare and Community Development in the late 1960s) was made from locally available scrap metal and hand-backed clay tile, brick and masonry, and was claimed to bring about a 50 per cent saving in fuel. By the mid-1970s, however, it was clear that many of the women who had earlier accepted the stove were not using it any longer, and that the experiment had been a failure.

Hoskins (1979b) on the basis of the women's complaints identified a number of reasons for the failure: the stove needed larger pieces of wood than were available locally; the stove surface was too high for stirring large pots; the sizes of the pot holes were not suitable for

*A traditional stove in Ghana: an improved design should also allow for easy stirring*

many of the pots in use in the house; if the unused holes were not tightly covered or the pot fitted only loosly, smoke escaped, the pots were dirtied, and more rather than less wood was used than before. In other words, the stove design was not suited to the users' needs as there had been little or no interaction between the designer and the women *prior* to designing the stove. A 'straight transfer' approach had been followed, where adaptation through user-interaction and involvement was necessary. Unlike the Guatemala experiment, however, no cases were noted here of the users themselves adapting the stoves.

Hoskins also provides some useful general insights into the sort of factors which have prevented the successful diffusion of wood-stoves in the many attempts to introduce them in African countries, such as:

> the failure to identify the key figures in the stove diffusion
> process, that is, the women who cook on the stoves, the local

artisans who can help in designing stoves and utensils and in taking care of repairs and alterations, and the local extension agents;

— the imposition of laboratory-tried models incorporating 'western' standards of improvement and ill-adapted to the local setting and cultural norms (including the length of cooking period, the time of day when cooking is done, etc.[5]);

— the failure to relate the physical elements of stove design to social realities. The adoption of such stoves often places additional burdens on the women: for example, in the Ghana study quoted earlier, the new stove needed larger pieces of wood, which meant going further afield—a trade-off which the women were not willing to make.

Both the Guatemala and Ghana experiments provide significant pointers. They strongly indicate that a close interaction between designers, users, local artisans and extension agents is likely to be a crucial element in the successful diffusion of wood-stoves.

This is further illustrated by an alternative, on-going experiment for diffusing wood-stoves in the villages of north-west India, where a highly personalised and participative approach to diffusion has been followed. Sarin, the principal initiator of this alternative experiment, in her *Chulha Album*, provides a detailed and fascinating account of the process by which she helped construct her first 29 stoves in the villages near the city of Chandigarh (Sarin, 1981). She describes how each user's problems were resolved and the specificities of their needs taken into account. It is apparent from the *Album* that many of the problems which cropped up could only have been taken care of and resolved through the close participative interaction with the users that characterised the building of the stoves. Each stove, built from local clay, was made user-specific in terms of its location within the kitchen, its size, the cooking routine of the family, the number of pot holes, the size of the pots and the overall aesthetics of design. The stove was usually built jointly by Sarin and the female members of the household, with other village women sometimes helping in or observing the process. Modifications were made after the user had utilized the stove for some time and found some aspects unsatisfactory.

This method of stove development and diffusion has been found to have several observable advantages:

(i) Since the stoves are closely adapted to user needs, user satisfaction is high, as expressed by 27 of the 29 households reviewed in the *Album*. Also, the flexibility of the approach is especially advantageous for taking care of the needs of poor households: some of these cook only one item and need a certain economy of design to maximize the use of available fuel heat; others have only thatched roofs and need special care for reducing the fire hazard, such as sealing the chimney with rags or clay to prevent its hot contact with the thatch, etc.[6]

(ii) As the women users are closely involved in building the stove and can thus gain familiarity with the basic principles involved: (a) they can carry out minor modifications and repairs themselves; (b) some can even build the stove on their own—one woman built herself a second stove, and another built a fairly successful first one on her own merely after watching it being constructed in a neighbouring house;[7] (c) the stoves are looked upon by the women (who often decorate them in patterns and colour) with pride, and as items to be shown to neighbours and relatives especially during festivals, weddings etc. This has been found to have a strong demonstration effect both within the village and in neighbouring villages from where requests for similar stoves often follow. In other words, there is an automatic propagation of the improved stoves in informal ways, which few official channels can easily duplicate.

(iii) Since the process of building usually involves close interaction with the household members it often helps to build up support for the stove from husbands or other men in the household. This too facilitates the acceptance and dissemination of the stove.

Initial evaluations of the stove indicate a high degree of success in terms of providing clean cooking conditions. In all the 29 households in the villages where the experiment was first conducted, women noted that the new stoves helped eliminate smoke and thus prevented the blackening of pots, walls and clothes and saved their time and effort in scrubbing and cleaning; it was also easier to start a fire. (Some men mentioned that it reduced their waiting time for meals, tea and bathing water due to the possibility of simultaneous heating and cooking.) Although no systematic tests were conducted on wood saving, 11 users mentioned that it did save wood.[8]

In another area, a woman worker who had received a six day intensive training (which included constructing a few new stoves, and repairing some old ones) under Sarin, built 30 stoves using a similar approach. A follow-up visit by Sarin indicated that the stoves were correctly constructed and functioning well; also, by the tests conducted by the woman worker, in all 30 cases there was fuel saving (Sarin and Winblad, 1983).

Of course questions about the replicability of a dissemination and training programme of the kind undertaken by Sarin, on a large scale, can legitimately be raised. But one can also query whether less painstaking approaches have worked.

Attempts at rapid dissemination, especially through inadequately trained persons, can even be counterproductive, as strikingly brought out by Sarin and Winblad (1983) in their report based on a visit to the *Safai Vidyalaya* (Sanitation Institute) in Ahmedabad (India). This organisation is reported to have been responsible for getting over 35,000 stoves installed in the villages of Gujarat. In terms of numbers alone, it is believed to be the largest dissemination effort in the country. But a closer examination of the programme has brought to light several less cheering aspects. The success rate of stoves (measured simply in terms of the per cent of stoves installed that were in continuous use) is estimated by the promoters (on the basis of informal feed-back) to be 50 per cent. The emphasis is on smoke removal rather than on energy conservation. And the 'campaign' approach (holding short, e.g. ten day, demonstration camps to motivate and involve the villagers, government functionaries and students, etc.) which is part of the organisation's dissemination strategy is found to have several limitations. For instance, in two villages visited by Sarin and Winblad, the stoves had been built by students who had attended a camp organised by the *Safai Vidyalaya* where the construction of smokeless stoves had been demonstrated. In both villages it was found that the stoves constructed were technically faulty; in many cases the chimneys ended within the hut so that all the smoke was released under the kitchen roof itself, and none reported fuel saving. By Sarin and Winblad's assessment, the inadequate training of the students (whose only exposure was watching the stove being constructed in the camp), their limited understanding of the principles of combustion and draught, their

*Sarin's participative experiment: teaching the women to build the stove themselves (plates A to D)*

A

B

C

urban background with no long-term involvement in the village, and the lack of a follow-up—all resulted in stoves being built that were neither smokeless nor fuel conserving.[9]

Such 'campaign' approaches are clearly not the answer.[10] Yet no ready answers are provided either by the existing State structures and extension networks through which rural innovations are usually disseminated—a point which will be taken up in detail further on.

Also, the degree to which improved wood-stoves, when adopted, can help save wood is still uncertain. Existing evaluations on this count are few, usually of an impressionistic nature or conducted on very small samples and, by and large, unsatisfactory. What is nevertheless clear is that laboratory promises of a 50 per cent saving of wood are rarely fulfilled in practice.

## 2. Economic Aspects

The private financial benefits of investing in an improved wood-stove as already noted, are likely to be small or nil where wood is still not generally purchased. The private financial cost of the investment would depend on what materials are used to build it. Where built from material available locally, such as local mud or clay, the expense may be negligible. Where the material is difficult to procure and needs purchasing, some financial expenditure would have to be incurred. In fact in several countries (such as Upper Volta, Niger and El Salvador) the improved stoves being promoted have to be purchased and often the costs, despite State subsidy, prove too high for poor households. In Upper Volta, for example, it is noted that all those purchasing the stoves are middle and upper-middle class households who buy the stoves basically to reduce smoke but have little incentive to save fuel (Foley and Moss, 1983: 112). There could also be an indirect financial cost if the stove necessitates the purchase of new cooking utensils. Among the non-financial benefits of investing in an improved stove would be the saving of women's labour time (in wood collecting, cooking, lighting fires, scrubbing blackened vessels, etc.), the absence of smoke (although, as noted, this may not always be seen as a benefit), the greater ease of cooking where the stove is adapted to the most comfortable cooking posture, the saving of cattle dung (currently burnt as fuel) which has an alternative use as manure, and being able to maintain or improve nutritional levels.

*Using dung for fuel instead of as manure in India: high opportunity cost*

Mark Edwards/Earthscan

From this list, it would be apparent that most of the potential benefits from improved wood-stoves are likely to be non-monetary and often in the form of intangibles, while the costs may in fact be monetary in nature. Also, these benefits may not always be perceived by the person making the decision to adopt, as for instance where the household men make the decisions and the benefits accrue mainly to the women. Further, the benefits, even as they stand, would not necessarily accrue to all the stove adopters. To begin with, the extent of non-financial benefits is in many ways dependent on the economic class of the household. For example, the saving of cattle dung for manure would only be important to a land-cultivating household and not to a landless one. At the same time, the effects on nutrition levels would essentially be felt by those households who are on the margin of subsistence, who cannot afford to buy alternative fuels and who therefore have to economize on it. And it is precisely in this case that it would be very difficult for the household to buy a stove, even at a very low cost.

All these aspects introduce complexities in wood-stove diffusion and once more highlight that the woodfuel problem is also a problem of poverty. Wood-stoves cannot, therefore, merely be placed on the market and promoted through advertisements, depending for their dissemination on the individual's ability to pay. The issue of adaptation apart, even where there is a felt need among the poor it would have to be made demand-effective, say by providing credit to the women users. Also, like contraceptive diffusion and health-related programmes, stove acceptance is determined by a range of factors in addition to the purely financial or economic.

### 3. Infrastructural Aspects

In the present context, public infrastructure may be seen to serve basically three functions in the diffusion of rural innovations.

— in the development of the innovation;
— in spreading knowledge of the innovation to the user: the provision of extension services;
— in making it feasible for the potential user to acquire the innovation: the provision of credit.

What appear at first sight to be fairly straightforward functions of providing physical facilities are in fact complex, since whether or not these facilities serve the needs they are set up to fulfil depends crucially on the approach followed in delivering these services, and on their degree of susceptibility to biases in favour of certain groups over others.

The issue of the appropriate approach to innovation development and extension was dealt with in detail earlier, when we noted how direct user-involvement can be a significant help in adoption, and that the line between innovating and diffusing can be a thin one. The focus here is thus on the biases in extension and credit services. Clearly, many of the issues discussed in this context would apply to most rural innovations, hence evidence on these issues in relation to other innovations would be of relevance here. Much of the available evidence on the bias in access to information on new innovations relates to the spread of agricultural technology, especially HYV cereals. In this context, it has been noted both in Asia and Africa that the mass media and extension agents tend to favour the economically and socially privileged households: village level agricultural extension workers, typically, contact the richer land-owning farmers.[11]

The elitist attitudes displayed by extension workers *vis-a-vis* the rural poor have also often been commented upon. Leonard (1977), for example, notes in his study relating to Kenya, that extension agents who had secondary school education performed less well than those who had only primary school education, both in terms of the number of visits they paid to the farmers, and in their level of knowledge and ability to communicate that knowledge. He attributes this (a) to their lack of motivation—being better educated they tend to think that the job is unworthy of their talents; and (b) to their having acquired, through their greater exposure to urban influences and values, elitist attitudes towards the rural areas and towards farming, which is considered by them to be an inferior occupation. Again, Lele (1975: 99) in the context of Africa notes that '. . . planners often feel that the peasant is incapable of adopting recommended changes on his own initiative and look on adult peasants as only big children who need constant watching'. Likewise, Bajracharya (1983a) on the basis of personal observation in Nepal notes that government officials are often condescending and at times even insulting towards villagers when they undertake village tours or when villagers visit government offices.

Further, extension services tend to favour men over women. In

*Typically, male extension agents speak only to the better-off male farmers: India*

both Asia and Africa the government extension agents are typically male[12] and generally contact the household men, even when the information is directly relevant to the women, as is, say, agriculture-related information to women farmers. The justification provided by some African extension personnel is that: 'In the African way we speak to the man who is the head of the house and assume he will pass on the information to other members (Staudt, 1976: 91). Therefore, women at best get second-hand information and at worst none at all.[13]

If the same biases carry over to programmes relating to improved wood-stoves, these are likely to adversely affect diffusion. For instance, if information is supplied only to men, women would not be in a favourable position to make or influence decisions on stove purchase. Further, appropriate adaptations would not be possible if women, who as users are in the best position to make suggestions, are not consulted. Also, given that in overall terms women tend to be isolated from the flow of technical information, this is likely to have adverse implications for the accumulation and development of indigenous technical knowledge and skills. (There is clearly a case here for recruiting women extension agents who would not have the same problems as men in gaining access to women users.).

*Directorate of Extension, India*

*Desirable but atypical: a woman extension agent speaks to women farmers in India*

In addition to the noted biases, a related factor is the work conditions of the extension workers. The village extension agent usually has to handle a range of issues from agricultural inputs to family planning, but the training s/he receives is often not adequate to provide up-to-date knowledge (IADP, 1966; Lele, 1975).[14] Further it has been observed that extension staff are very frequently transferred, so that any local experience acquired or rapport established with the villagers, cannot be used to full advantage.[15]

The issues are thus two-fold:

— the method of bureaucratic functioning in Third World countries, which relates to the aspects of training, transfers, etc., and essentially determines the quality of information imparted through the extension system;
— the bias in approach and attitudes of the extension agents which determines to whom the information is imparted.

Some initial evaluation of recent attempts to diffuse stoves through government extension channels in India indicates that these can be significant limitations. A report assessing a programme to spread pre-fabricated stoves among rural households in Gujarat notes that it has made little progress. The reasons identified are the lack of commitment by various agencies involved in stove programmes, a lack of women extension workers, poor performance of stove design and a lack of adequate contact of the community forestry wing staff with household women (World Bank, 1983a). An additional likely factor which the report does not stress is the cost of the stove (whicn is Rs.50). A programme to promote improved stoves in Uttar Pradesh (India) has fared even more poorly. Since 1980 when the programme was initiated only *one* stove has been distributed (World Bank, 1983b).

In some recent attempts at wood-stove diffusion in countries such as Senegal, Kenya and Sri Lanka, the effort has been to train local potters in stove building and also to involve women both in extension work (e.g. by training them to build stoves) and as users (see Joseph, 1983). These are clearly steps in the right direction, but more detailed evaluatory work is needed to indicate the exact method of functioning of such items in practice, and their success in diffusion.

Next comes the issue of credit availability. This assumes importance to the extent that the stove itself, or the materials for building it, need to be purchased. Of course in absolute terms the expense would be small. But given that many of the potential stove users are the very poor who often have to incur debts even for consumption purposes, any extra spending, even if small, could depend crucially on access to subsidized credit for this purpose. In El Salvador, for example, one of the significant difficulties in stove diffusion was noted to lie in the fact that the stoves cost US $18 each, while almost all household heads were women with no paid employment and hence no cash to buy the stove (Sandoval, 1983). Likewise, in an evaluation of two wood-stove diffusion projects in the Sahel, a lack of liquid cash for stove purchase in the hands of the women (especially of poor households) was identified as one of the important constraints to stove dissemination; and it was emphasised that a credit scheme for the women would help (Norman, 1981).

However, just as there is a bias in access to information against the underprivileged groups in the community, so there is a bias in access to credit. There is overwhelming evidence for both Asia and Africa that cooperatives and other rural credit institutions typically tend to be monopolized by the economically and politically powerful groups in the rural community.[16] Further, women are in a worse position than the men (even in those households where men can get credit), since access often requires guarantees in the form of land titles which they seldom possess (Staudt, 1976: 87-88). Hence they would usually not be in a position to obtain credit independently of their husbands. Also, institutional credit is generally given only for productive investments, and insofar as wood-stoves count as consumption items, specific schemes for this purpose would need to be introduced.

## 4. Cultural Aspects: Attitudes to Change

Not infrequently, problems of rural diffusion are attributed to the 'irrational', 'conservative' attitudes of rural people. Such explanations can be misleading. More often than not, the problem is located in the potential adopter's particular economic and social position within the community. Further, what may appear to be irrational to an 'outsider' may in fact be perfectly logical within the potential adopter's cultural context, and an understanding of this context would be crucial for successful diffusion.[17]

An interesting illustration is provided by Bajracharya's (1981) study of firewood use in Nepal, where he notes how one set of his sample households use wood-stoves and hence less wood than the rest who use the open fire. These different technologies co-exist even though the households are located close to one another and the existence of efficient wood-stoves is common knowledge.

Here one or both of the following explanations would be valid. The first relates to religious beliefs and rituals. The households using the open fire believe that the *pitri devta* or 'family spirit' resides in it, and their reluctance to switch to the stove could be attributed to superstition. They belong to the indigenous caste groups in the area—the *Rais, Gurungs*, etc. The stove-using households have migrated from outside, albeit some generations ago. They belong to the *Brahmin* and *Chhetri* castes and have a somewhat different set of religious customs. A second explanation relates to the fact that drinking alcohol is common among the *Rai/Gurung* communities who brew their own liqour. This is done in large pots which need the wider open fire rather than the narrower wood-stoves; among the *Brahmins/Chhetris* liquor is less commonly consumed.

Both explanations emerge from the particular cultural milieu within which diffusion is being attempted. Insofar as the cause of non-adoption lies in drinking habits, stoves could readily be designed to take this into account, although adaptation to take account of religious beliefs is more difficult. Yet even here some types of religious beliefs can easily be accommodated: for instance in certain parts of India, certain directions are considered inauspicious and stoves facing those directions would not be used; but if the stove promoter is familiar with local custom, the stove can be built appropriately (Sarin, 1983). Also, beliefs have themselves been known to be adaptable: for instance, the three-stone stove is considered a symbol of a united family in African communities, but this has not prevented the women from using alternative fireplaces for particular uses, or from adopting the stove in various ways, such as by putting an extra rock or two between the stones on a windy day (Hoskins, 1979b: 41).

The potential adopter's attitudes towards particular innovations would also be governed by the person's past experience with innovations; and, equally important, the person's experience with past promoters of innovations. Where the same set of extension agents are used for promoting wood-stoves, as are used for promoting a range of other rural technologies from HYVs and mechanical equipment to contraceptives, their credibility with the potential adopter would depend significantly on the degree of success with these other innovations. This is because the adoption of any new innovation contains an element of risk for the user. It might be a risk of financial loss as involved in switching from traditional seed varieties to HYVs, or a risk of inconvenience if say a new stove does not work as expected. An innovation well adapted to user's needs and economically cheap would no doubt minimize the element of risk, but even where the risk is *objectively* low, the user may still need to be convinced of this, and the ability of the innovation-promoter to do so would depend on the success of her/his efforts with innovations introduced previously.

In some cases it is easier to get acceptance for a completely novel item compared to modifications in an old one. For example, Joseph (1980) notes how among the oceanic people who were unused to cooking pots, there was ready acceptance of cooking pots with lids, while in other communities where lid-less pots were common, the tendency was to remove the lids of the new pots as well.

Basically, the above discussion reinforces the need for those promoting the innovation to have a deeper understanding of the way in which the community, of which the potential adopter is a part, operates; it requires an insight into the complex set of factors that govern behaviour and provide an overt or covert logic for doing a certain thing in a certain way.

Such an understanding cannot be gained in the laboratory—it necessitates a closer interaction with the potential user. Hence, once more, the issue of innovation adaptation in the field and of user-involvement emerges as being one of significance.

## 5. Social Structure: The Link

We now come to the linking issue of social structure. Here the attempt will be to indicate how inequalities in social status and the unequal nature of power balances between different classes/castes etc., and between the sexes are likely to affect wood-stove diffusion. The threads to this have already been provided in the discussion so far. Here it is hoped to show how they interweave.

Consider first the question of women's status. The importance of this had earlier been touched upon only briefly, and the issue needs some elaboration. The status of women within the household could be a significant factor in wood-stove adoption, especially where adoption requires cash expenditure, by virtue of the fact that although women are the potential users of the innovation and therefore in the best position to assess its advantages and disadvantages, it is men who usually handle the household cash and make decisions on how it is to be spent.[18] Generally, rural women and men are noted to have differing priorities in household expenditure. For instance, in poor rural households if women have some independent earnings they are observed to spend them on the family's basic needs, while the men spend their earnings largely on their own needs, such as liquor, clothes etc.[19] Hence where men make the decisions, the purchase of an improved stove may not get priority, especially where the only advantage perceived is greater leisure or convenience in cooking for the women. It is in fact noteworthy that in the heart of India's green revolution, viz. Punjab, while there have been considerable improvements in the technology handled by men, in the form of tractors, threshers, combines, etc., there has been little improvement in the women's kitchen apparatus, even in the economically well-off families (Agarwal, 1984: A-49).[20]

*An open fire in Sind, Pakistan*

Likewise, the status of women within the community enters as an influencing factor in a number of significant ways. As noted, rural women usually have no direct access to institutional credit or to independently disposable cash income to purchase new innovations/technologies. This has been found to constrain dissemination in several wood-stove diffusion programmes as was noted, for instance, in the Sahel and El Salvador.

Further, women seldom have access to information on new innovations. There is too a strong ideological bias in extension services which is likely to work against the direct involvement of, or consultation with, village women in the experimental designing of wood-stoves for their use—an involvement which was found to be a significant feature in effective diffusion in several case studies, such as those relating to Guatemala, Ghana and India. Also, rural women are not usually given the training or the opportunity to undertake decision-making roles or responsibilities in the public sphere, which would be necessary to undertake in extension work.

Consider next the issue of the balance of power between rural households which differ in their ownership and control of material assets and/or in their social status. (Political power and social status generally, though not always, coexist with the ownership of wealth.) This affects the ability of different households to purchase (with their own incomes) technologies which require financial expenditure and also affects their access to information and to credit.

Further, social hierarchies, whether based on differences in inter-household distribution of material assets, or on gender differences, or defined by some other criteria, are likely to make difficult the setting up of precisely those linkages between indigenous technical knowledge and skills and the more formalized research and development networks, and between the user (including the user-innovator) and the scientist/professional, that were identified earlier as being important in the successful diffusion of rural innovations in general, and wood-stoves in particular.

That indigenous technical knowledge and capability does exist in the rural areas of Third World countries can be supported by several examples of innovations, especially in the context of agricultural technologies. For instance, Dommen (1975)[21] documents the development of cheap bamboo tubewells by farmers in Bihar (India). Sansom (1969) describes the invention of a small centrifugal motor pump by two Vietnamese farmers who had only on-the-job knowledge (from having worked for a while on French dredges) and no formal training in engineering, but who developed two similar versions of the pump independently of one another. This pump was rapidly diffused, despite the absence of governmental encouragement and the dependence on informal communication channels. The introduction of improved agricultural practices by farmers has been noted by Hayami and Ruttan (1971: 157) for Meiji Japan, and by Howes and Chambers (1979: 6) for Nigeria. Biggs (1980) refers to several cases of farmers in India and Bangladesh bringing about genetic improvements in crops.[22] Improvements in stove designs by local people have already been described in the Shaller study. Dutt (1978c: 6) provides another example of local skills in the context of handpumps and wood-stoves in India. Here a night-watchman at the

Ungra Extension Centre in a Karnataka village, became an expert in handpump servicing merely by observing various technicians at work, and provided the service to many villagers in the area. Also, as a follow-up on some tests carried out on the smokeless 'HERL' stove, he undertook the task of recording, over an extended period of time, the weight of wood used for cooking. This villager's training was entirely unplanned.

That such innovations and innovative skills usually remain isolated instances and are rarely picked up and integrated into the formalized research and extension systems, points essentially to the weakness of existing links between the rural user-innovator, the extension worker and the scientist/professional. And it is precisely the strength of these links which is likely to determine the efficacy of attempts at field adaptation and diffusion of wood-stoves (or any other innovation with similar characteristics) on a mass scale.

Typically, there is an absence of a two-way interaction—a dialogue—between the scientists/professionals, the village extension agents and the poor peasants or other underprivileged (in particular women) users of innovations.[23] The bias of government extension workers who enjoy a certain status in the village as part of a well-entrenched bureaucratic hierarchy (even if they may be at the lowest rung of that hierarchy) in favour of rich landowners and against the poor peasants, has already been noted. But the problem is only partly one of economic inequalities. Underlying the divide between the scientists/professionals (usually urban-based) and the rural users of innovations (including user-innovators) whose knowledge comes more from field experience than from formal education, for example, is also usually the divide between mental and physical labour, between town and countryside, and between the sexes.

This is not to say that examples of localized experiments to establish these links, manifesting a 'participative' (of the rural poor) rather than a top-down approach to diffusion, do not exist. In fact there are several micro-level success stories which serve to demonstrate, on the positive side, what can be achieved through a dialogue between those seeking to diffuse innovations or innovative ideas and those whose lives are directly affected by the programme. At the same time, they also point to the need for more comprehensive social changes if these experiments are to become general and wide-based. One such experiment in the case of wood-stoves in India has already been noted (Sarin, 1981). Two other illustrative examples from India are the Banki piped water supply project in Uttar Pradesh (UP) and the Jamkhed Community Health Project in Maharashtra. Both these examples are of special interest since they bring out the contrast between the ineffectiveness of the conventional approach to the delivery of these services and the effectiveness of a participatory approach. Also, these experiences are of specific interest in the context of wood-stove diffusion, because the basic characteristics of public health care, rural piped water supply and wood-stove diffusion programmes are similar in many respects. For instance, they all provide mainly non-financial benefits to the users, and usually involve little private financial cost; for effectiveness they need to cover a large section of the underprivileged population; and the primary persons who need to be incorporated in each are the rural women who bear the main burden of family illnesses and of fetching water and firewood.

In the Banki project (Misra, 1975) the 'innovation' that was sought to be introduced was piped drinking water, to a population which until then had used open wells. Initial attempts made in the 1950s by the State Irrigation Department, which supplied tubewell water to the fields, to construct overhead tanks in some villages, lay pipes and provide public standposts, met with failure. A survey in 1962 revealed that 96.1 per cent of the population in these villages still used open wells. Little attempt had been made to involve the local people in project planning and implementation. The approach of the Banki project, which was started in 1962 in seven of these villages by the UP Planning Research and Action Institute (and funded mainly by UNICEF and WHO, with land and small sums being contributed by the *panchayat*[24] head and some villagers), was in sharp contrast to the Irrigation Department's top-down method. It aimed in fact at developing a scheme where the people would have their own installations and would ultimately be able to administer the system independently. It was sought initially, through a base-line survey, to understand why people were reluctant to accept the innovation. Some of the apprehensions expressed related to the tastelessness of tap water, the possible harmful health effect of drinking electrically pumped water, anticipated water charges, fears that the water was medicated to reduce fertility, etc. In order to remove misconceptions and create a positive attitude towards the scheme, discussions were organized with the villagers in informal evening 'sittings'. To demonstrate the importance of clean drinking water, a health education programme was simultaneously started. By 1966, a substantial proportion of the households was using piped water, and by 1973 all the families were doing so. Over a third had taken private home connections. The villagers had also assumed full responsibility for the management and general maintenance of the system, through its Waterworks Executive Committee composed of seven villagers (one from each village) which was recognized by the State Government.

In the Jamkhed Community Rural Health Project initiated in 1970 by two doctors (a husband-and-wife team), the emphasis again has been on community participation in decision-making, with the ultimate objective that the villagers will run the programme themselves (see Sethi, 1980, and Malgavkar, 1981). Till then, the services available in this drought-prone area of Maharashtra had been inadequate (a dearth of doctors willing to go to the villagers) and costly (high costs of doctors' fees, medicines prescribed, travel, etc.). The project has sought to provide an alternative, low-cost health service through a team of locally trained paramedics working with the doctor-couple, with a referral system, in case of need, to other doctors. Local women (whose names have usually been suggested by the villagers themselves in group meetings) have been trained, initially as auxiliary nurses-cum-midwives and later as village health workers to promote curative and preventive health care. Although the majority of these women are illiterate, they are quick to learn, and their ability to communicate with and gain the confidence of the other village women (helped by commonality of diction, tradition and values) has been one of the main strengths of the programme. There is a conscious attempt to overcome caste barriers (the paramedics come from all castes and have to attend to the needs of all castes), and to maintain a relationship of equality between the professionals and non-professionals.

The results have been impressive in terms both of health statistics and of developing self-reliance and organisational ability among the poor. Initially begun in eight villages, the project has now spread to over seventy. It is noteworthy that one of the main impediments in programme implementation is reported to be the attitude of the town doctors who 'feel that a decentralized health care service offered through paramedics will reduce their practice and affect their income' (Malgavkar, 1981).

The experiments described provide a noteworthy demonstration of the effectiveness and necessity of an alternative approach to diffusion. They also represent significant beginnings. At the same time, it cannot be ignored that, at present, they remain micro in nature in terms of the percentage of total population covered, and exceptions relative to the large number of experiments/projects which continue to rely on the top-down approach. Underlying their exceptional nature, however, is precisely the difficulty of operating such programmes within well-entrenched, hierarchical social structures.

## Notes

1. In any case, even when women do wage work they are still usually responsible for collecting firewood for household needs (e.g. see Fleuret and Fleuret, 1978).

2. In the 'top-down', straight transfer approach, the development and the diffusion of the innovation are seen as separated in time and space; with field adaptation and user-involvement the two aspects, in many ways, go together.

3. An understanding of user-specific needs becomes even more necessary when we consider that within a given region there may be several culturally distinct, indigenous communities, as Shaller (1979: 3), for instance, notes exist in highland Guatemala.

4. An attempt to diffuse the Lorena stove in El Salvador similarly had limited success (Sandoval, 1983).

5. Hoskins (1979b: 37) notes how 'specialists' frequently assume that women can easily cook in the daytime instead of the evening without noticing that during the day women are working in the fields to grow the grain to fill the pot.

6. These are described at some length by Sarin in her stove-makers' manual (see Sarin, 1983).

7. There can also be other positive results of such a participatory approach to diffusion which are not immediately observable. For instance, in Senegal, where the Ban-ak-Suuf stove was promoted by involving women potters it was noted by an experienced field worker of the programme that:
   'The positive aspects of participation in a group activity such as stove building can have a tremendous impact on village women. Most women are amazed that men's participation in the actual 'main d'oeuvre' isn't obligatory and that they can do it on their own. Even if Ban-ak-Suuf stoves have somewhat of a short life expectancy, all the preliminary work, organisation, and participation that leads to the actual construction is an accomplishment worth noting in terms of community development. Women can realise a goal in a new domain and the skills they acquire in doing it can then be used in the future' (quoted in Foley and Moss, 1983: 111).

8. In one case a husband claimed that the bundle of firewood which earlier lasted 2-3 days now lasts 7-8 days.

9. Sarin and Winblad (1983) provide an interesting critical survey of several wood-stove dissemination programmes across the country based on their visits to these

areas. In most cases they note inadequacies in stove construction or in methods of dissemination used.

10. This example also highlights the problem inherent in seeing the designing process as totally distinct from the dissemination process, in the case of innovations such as wood-stoves.

11. See, e.g. Dasgupta (1977), and Griffin (1971) on the Asian experience; Leonard (1977) on Kenya; and Lele (1975) and Hapgood (1965) on the African experience in general.

12. Among possible exceptions would be health services where women are sometimes trained for giving contraceptive advice to rural women. However, other government health workers such as those responsible for innoculation/vaccination campaigns, etc. are again usually men. Non-governmental voluntary organisations working in the field of community health also sometimes seek to train local women as paramedics (e.g. see Malgavkar, 1981, on the Jamkhed experiment in Maharashtra, India), but again these would constitute the exceptions and not the rule.

13. Lele (1975: 76-78) also speaks of the neglect of African women in extension services. In African countries, in particular, women have long been farmers in their own right in many areas—see, e.g. Boserup (1970) and Agarwal (1981).

14. In relation to information on agricultural technology this has been noted to lead to a 'widespread lack of confidence and reliance amongst cultivators on extension officers for HYVP (high yielding variety programme) guidance' (PEO/AMU, 1976: 11). Of course this lack of confidence can also result where the agent does not practise what he preaches. For example, in Kenya, some agents are also farmers of small plots of land but they do not always follow the agricultural practices they recommend to others (see Leonard, 1977: 16).

15. In the course of the Intensive Agricultural Development Programme (IADP) launched in India in the early 1960s, it was noted, for example, that 40-50 per cent of the IADP staff from the Block Development Officer downwards, served a tenure of less than two years and often of less than one year in any given area.

16. See Dasgupta (1977: 115-116) for India, Apthorpe (1970) for Africa, and Dumont (1973) on credit cooperatives in Bangladesh.

17. The myth of the conservative, non-innovative, Asian farmer (Lewis, 1951) for example, has by now been exploded successfully in the face of ample evidence that what has often prevented the poor peasants from innovating has been their economic and social situation, which makes it difficult for them to take risks or gain access to new inputs (Schultz, 1964; Mellor, 1966; Wharton, 1971; Lipton, 1968; Hapgood, 1965).

Hapgood (1965: 9) in the context of Africa puts it succinctly: 'The farmers, they (many Africans and foreign technicians) tell us, are too stubbornly opposed to change to accept innovations that would benefit them. This is a convenient way of shifting the blame from the shoulders of those who planned an unsuccessful project to the farmers who have rejected it. We rarely hear the farmers' side of the story. Nor does this explanation fit the known behaviour of African farmers. They have shown themselves quite ready to innovate over the past century, having quickly adopted cash crops such as cocoa and peanuts.'

18. Here we would expect differences between African and Asian households. In the former, women are often cultivators and/or traders in their own right and are

more likely to have some independent access to cash, compared with Asian women who rarely cultivate plots separately from the male members. However, in practice, the difference might be marginal, since even in the African context, the cash is still often controlled by the man (see e.g. Bukh, 1979: 29).

19. This is especially highlighted in the Hanger and Moris (1973) study which relates to the effects of an irrigated rice resettlement scheme in Kenya. Under the scheme the incomes of the men increased substantially but little of this increase was shared with the women. The household expenses also increased meanwhile—e.g. in the old villages firewood could be collected by the women from communal land; in the scheme villages it had to be purchased. The money that the women received from their husbands for such expenses was rarely sufficient to provide adequate firewood. To reduce firewood consumption women often lit only one fire, cooking with only one pot. Also see Bukh (1979: 29, 54) for Ghana, Arens and Van Beurden (1977: 45) for Bangladesh, and *Consortium of International Development, Vol. III* (1978: A-53) for several other countries, on the divergence of male and female expenditure patterns.

20. More recent efforts to bring about improvements are largely initiated from outside, and often virtually costless in financial terms: e.g. the Sarin experiment described earlier.

21. Also see Clay (1980).

22. Also see Biggs and Clay (1981).

23. An illustrative example of the gap which often exists between the needs of the rural poor and the perceptions of scientists/professionals as regards those needs, is provided by Herrera (1981: 33-34) in the context of a rural housing project, undertaken in South India, which turned out to be 'an expensive failure'. He also makes the general point that 'the scientist is working with problems that belong to his own economic, social and cultural background so he has a tendency to apply the same criteria to a completely different environment. He frequently assumes that he has to satisfy the same needs but on a lower level, due to limitations posed by the local economic conditions.'

24. The village council; usually a five-member elected body.

# CHAPTER V

# OF SOCIAL FORESTRY AND OTHER TREE-PLANTING SCHEMES

*Do not axe these oaks and pines—*
*nurture them, protect them,*
*From these trees the streams get their water*
*and the fields their green.*
*Look how the rhododendron smiles in the forest. . .*

*Wherever you see a vacant space plant trees—*
*fodder trees, oak trees, broad-leafed trees. . . .*

—from two Garhwali folk songs
by Ghan Shyam Shailani

'Social forestry' has today become the popular catch-all phrase for a variety of tree-planting schemes being promoted by governments and international funding agencies as *the* solution for both the woodfuel crisis and the environmental crisis. Yet an examination of these schemes indicates that few have moved any distance towards alleviating the crisis of woodfuel shortages faced by the poor. In this chapter the factors underlying the problems with these schemes are examined.

Tree production can be promoted under individual, government or communal management, on private, government or communal (belonging jointly to the village or community) land, and for different purposes—commercial or non-commercial. 'Farm forestry' usually implies individuals growing trees (to sell or for own use) on private land. 'Social forestry' on the other hand implies the planting of trees for meeting the needs (especially of fuel and fodder) of the rural people,[1] usually through the use of government or communal land, under government or communal management. Such tree planting when undertaken by the community (typically on communal land) is also often termed 'community forestry'.

In the last few years, a large number of schemes to promote tree planting under all three systems of management (but especially government and communal management), and on different types of land, have been launched—many funded by international aid-giving agencies, such as FAO, SIDA, World Bank, USAID. Yet available studies (some detailed, some impressionistic) evaluating these projects, indicate that while farm forestry has had a limited degree of success in some parts of the Third World (e.g. Gujarat in India), social forestry with very few exceptions has been a failure in most countries, even in the sense of ensuring the planting and maturing of trees, let alone in providing for the daily needs of fuel, fodder, etc. of the rural poor.

In many instances, the schemes have been actively opposed by the people. In Niger, villagers are noted to have pulled down trees planted under a World Bank project and allowed their cattle to graze uncontrolled in the area (Hoskins, 1979b). In Senegal, in one project, saplings distributed by foresters to the local residents for planting on forestry lands, were purposely destroyed by the residents (Hoskins, 1979b). In the Jharkhand area of Bihar (India) villagers cut down teak trees planted by the government in the local forest (Makhijani, 1979). In Tamil Nadu (India) 5000 hybrid eucalyptus saplings planted by the government were uprooted by the villagers (Subramaniam, 1983). In Ethiopia, landless labourers, when given saplings to plant, planted them upside down (Eckholm, 1979). In the Philippines, the Tinggian tribal communities are said to be starting forest fires deliberately (Aguilar, 1982).

In other instances, a passive resistance to schemes is apparent in the poor success of most government-promoted, community self-help projects. In India, for example, a mid-term appraisal of two large-scale, World Bank-aided forestry projects begun around 1980, in Gujarat and Uttar Pradesh, showed that while targets for farm forestry had been far surpassed, those for self-help village woodlots had fallen drastically behind (see World Bank, 1983a, 1983b). Reports of such failures are in fact numerous. How do we account for them?

The reasons must be sought in the range of complex, interlinked factors which emerge through a closer examination of the socio-economic context within which the programmes have been introduced and the methods of initiation and implementation of the schemes.

Consider first the context where there has been outright hostility to the government programmes. Many of these have been schemes where the government has directly taken charge of tree planting, usually through the forestry department.

## 1. Tree Planting by the Government

Some argue that the failure of such projects may be traced to there being no 'culture' of forestry as a productive activity among the people; that forests are generally looked upon with hostility or as impediments to be cleared or burned in order that land can be cropped as, for example, under the practice of 'slash-and-burn' agriculture. However, this explanation cannot be sustained in the face of existing evidence which establishes that rural people usually, and forest communities in particular, have traditionally shown a deep sensitivity to the need to maintain the ecological balance and to preserve rather than destroy. In the Philippines, for example, the Tinggians who are now said to be starting forest fires had earlier practised methods of forest conservation. They had constructed fire lines to prevent forest fires from spreading, and tree cutting in the watershed had been disallowed because of their recognition of the relationship between trees and water supply. In the early 1970s, however, their land was taken over by a commercial timber company.

Similarly, the often-mentioned practice of 'slash-and-burn' agriculture, as traditionally practised by forest communities in India and elsewhere, was not disruptive of the ecological balance—the land being left fallow long enough to restore vegetation. If today such practices are leading to ecological damage, this may be attributed (as noted earlier) to these people being pushed to depend on smaller and smaller areas of land, while large tracts of forest land have been taken over by others for different uses. Again, Gupta and Deshbandhu (1979) note how many tribal communities in Manipur (India) voluntarily regulate the commercial utilization of forest products. Land is set apart and trees on that land are cut in a planned manner for building public facilities, such as schools etc. The *Chipko Andolan* (tree-hugging Movement) to save trees in the Himalayan hills of Uttar Pradesh in India (discussed in detail further on) again points to the close relationship between people and trees. The scriptures and old texts in India and elsewhere are replete with sayings emphasizing the sacredness and importance of trees in people's lives. As Swaminathan (1982a: 339) who has been working with the tribals in Rajasthan for several years, emphasizes:

> 'There is nothing we can teach the tribals about trees. They know that rain falls heaviest where there are more trees. They know that soil erosion and the run-off increase with deforestation. They know the ecological changes that happen with hills being denuded of forests, how the soil becomes dry, sterile, hard and unyielding.' The forests are 'intricately woven into their culture, religion, songs, dances and rituals'.

For many tribal communities forests are their very life. Clearly, the causes of hostility to government programmes need to be sought in other than the presumed hostility between people and trees. Some of the likely causes are discussed below.

*Surrounded by forest products in Mali: for many communities forests are their life*

**(i)** *Alienation of People from their Traditional Rights to Land*

Many of the schemes where the governments have directly taken charge of tree planting have been introduced in opposition to the wishes of the people rather than with the support of the people. This is especially so where the schemes have involved the reservation of land for tree planting which restricts or terminates the rights of farmers, cattle grazers and hunters to the hitherto 'free' produce of the forest, and where the restrictions are imposed from above without the involvement of the local community in the decision. Such reservations have particularly hit tribal communities, whose main sustenance has come from the forests for generations.[2]

In some cases, such takeover of land has been to plant trees solely for commercial purposes. There are several examples of this from India. In the Jharkhand region (Bihar State) the government's forest department began a project to start a monoculture teak tree plantation, by cutting down the original mixed forests which had a variety of trees, including the sal that gave the people many types of produce. Not only were the teak trees being planted of no immediate use to the people, but in the process of establishing the plantation the forest department also deprived the tribals of rights to collect existing minor forest produce. As a result, the local residents have been axing down some of the government-planted teak trees, to deter the extension of the teak plantations at the cost of their sal forest (Makhijani, 1979: 39).[3] Again, in the Midnapore district of West Bengal, eucalyptus and other commercial varieties are noted to have been forcibly planted by the forest department on plots where the tribals originally grew paddy (*Indian Express*, 1983). In Uttar Pradesh, *sheesham* and sal were cut to plant eucalyptus (Dogra, 1984). Likewise in the Bastar district of Madhya Pradesh under a World Bank-funded project, 40,000 ha of deciduous forest was to be clear-felled to plant tropical pine as raw material for the paper industry. This brought strong resistance from the tribals, and led to a public controversy and the eventual shelving of the scheme (Guha, 1983; D'Monté, 1982).[4]

Even where the planting is ostensibly for eventual community use, the lack of an assurance that the people will benefit and the lack of an attempt to involve the people has led to scheme failure. For example it is noted that in Maridi (Niger), in a government-implemented rural development scheme, a project to establish 500 ha of village woodlots was incorporated. But the local people were not involved in formulating the project. The land on which the trees were planted was traditional grazing ground, access to which was precluded by the project. Hence people pulled down the trees and allowed their cattle to graze uncontrolled in the area (Hoskins, 1979b: 6, 7).

Again, in Madhya Pradesh (India), in a village in Rajgarh district, the forest department fenced off 65 ha of forest land for a forestry programme and another large tract for the establishment of a game sanctuary. The area fenced off for the forestry project had been used by the villagers for grazing livestock, and that taken for the game sanctuary contained a water pond for the livestock, a large space customarily used for the annual village fair, and stone quarries in which many villagers had been employed. The community was not consulted or made a party to the decision-making process, nor told that they would get the benefits of the community forestry scheme, in the long run. The result has been extreme hostility towards the scheme from the villagers, and the illegal felling of young trees (Sarin, 1980).

In general, the approach of the forest department in scheme implementation has come in for severe criticism in many countries.

## (ii) *Approach of the Forest Department*

In the context of Trinidad, Bernales and Vega (1982: 49) point out: 'It is noticeable that despite the concept of social forestry, foresters still tend to think ... solely in terms of trees and not of the community as well. This may partly explain why efforts at community organisation and welfare improvement appear limited.' Roy (1980) makes a similar observation in the context of India and describes how differences in the approaches of the forestry department and a voluntary agency both working on a forestry project in Ranchi district (Bihar State) brought the two into conflict. He observes that traditionally the forestry department has been tree-oriented and not people-oriented; and although now the officials accept that community forestry is 'for the people' they are still far from saying, as the voluntary agency emphasizes, that it is 'by the people'. There is a tendency too for the forestry officials to 'decide what we should do for them' while the voluntary agency approach is 'the community decides what is good for them and we help them in doing it' (Roy, 1980: 57). In general, Roy mentions that there is a feeling that the government projects make many promises but never fulfil them. According to him the tribal communities view the forest department officials as 'instruments of the government (which is distant and threatening)', and as exploiters who take bribes, harass, threaten and extort in the name of unfamiliar laws, and who are gradually taking away from them their natural habitat, namely the forest.

This feeling is usually based on long experience. Guha (1983), for example, historically traces the role of the forest department in India. He notes how the creation of the department under colonial rule in 1864 was accompanied by legislation curtailing the previously unlimited rights of users over forest produce and giving the State monopoly rights over land. Also, through various forest acts it sought to establish that the customary use of forests by the villagers was based not on 'right' but on 'privilege', and this privilege could only be exercised at the discretion of the local rulers. Large tracts were also demarcated and fenced off, increasing the pressure of the population on the remaining land. Not surprisingly, this period was marked by many revolts by forest communities. Also, as a superintendent of forests in Dehradun in 1897 (quoted in Guha 1983: 1885) mentions: 'not altogether without reason, the villagers believe that any self-denial or trouble they may exercise in preserving or improving their third class forests, will end in appropriation of the forests by the forest department as soon as they become commercially viable'.

Yet, in many ways, the attitudes and method of functioning of the department have not changed substantially. Romm (1979) notes how in Madhya Pradesh the government has legal claims to teak and *sisoo* trees on private land and mentions that there are numerous examples of 'social forests' in which villagers have seen the government come at will to harvest and sell the produce. Sarin (1980) gives specific instances of villages where fruit trees planted by villagers on their farms have been claimed by the government as its property. Now the villagers refuse to plant trees without a written assurance that the trees planted on their own land are their property.

Several people have also commented that the forest department is often directed by self-interest and not the interest of the people. For instance, Verma (1983) commenting on a draft social forestry project in Rajasthan which has been prepared to attract World Bank aid, notes that a significant part of the envisaged expenditure is to be on barbed wire, and personnel: out of a total budget of Rs.51 crores, 8 crores has been demarcated for barbed wire fencing and 12 crores for establishment salaries. He ironically remarks:

> 'At the end of the project period all that the people of Rajasthan will have from the social forestry project is 50,000 tons of corroded barbed wire stretched on termite eaten bullies, across the heart of Rajasthan for over 100,000 km. The State will be poorer by a few hundred thousand trees cut to make the uprights for barbed wire fencing, and saddled with the permanent liability of 2403 forest officers and guards costing Rs.2 crores annually to the exchequer.'

He notes bluntly: 'The fact is that the forest department is not fond of people', and points out that the department has not even sought to talk *to* the people let alone *with* them.

(iii) *Attitudes and Practices of Forest Officials*

> *The foresters themselves are getting the*
> *forests felled;*
> *making profits from the trees they sell.*
> *What if the fence itself wrecks the field?*
> *If the water itself ignites the fire?*

—from a Garhwali song sung by *Chipko* activists[5]

Problems with and even hostility towards government programmes stem not only from the land use policy of the forest department but also from the attitudes and practices of the forest officials, especially the forest guards. These practices range from the elitist approach of the guards to outright exploitation.

Hardjosoediro (1979), for example, locates the failure of an experiment to promote teak planting along with crops in Java to the failure of the forest officials to establish a close interpersonal relationship with the peasants and the local population. The scheme he describes was conceived by the forestry service as a system of shifting cultivation—*taungya*—under which peasants were contracted to plant and tend teak saplings usually for two successive years. In return they got a small fee and had the right to grow crops in between the trees. A foreman was appointed to supervise the work assisted by four volunteers—usually elders selected from amongst the contracted farmers. Each foreman managed 10 to 12 hectares (covering about 40 to 48 peasant families). In practice the scheme has been facing problems. The peasants overcrop their small plots, neglecting the teak saplings. There have also been supervision difficulties, since, with the growing population pressure, the number of peasants contracted has been increasing and the size of plots allocated has been decreasing. At the end of the two years when the peasants have moved out (presumably the trees are by then too high to allow crop cultivation) there has often been uncontrolled cattle grazing; and under these conditions a large proportion of the newly planted land has been noted to revert to clear ground within five years.[6]

Hardjosoediro (1979: 114) feels that if the forest guards had been more accessible to the peasants, the latter's attitudes would have been different: 'A good forest guard in their eyes is he who is visible at any time and with whom they can share their luck and sorrow as their good neighbour, and from whom they can borrow food in time of emergency. . . .' He notes that in rare cases where the project officer stayed on the spot and established a personal relationship with the peasants, and where local labour was drawn into the project, there was an increase in cooperation from the local people, and cattle grazing or wood stealing in the project area was considerably curtailed.

However, this vision of forest guards is far removed from the reality of the relationship between forest guards and people in many countries, especially in South Asia.

Documentation from all over South Asia gives evidence of the exploitation of the tribals by the forest guards (who are usually placed at the lowest rung of the forest service hierarchy). There are several examples from India. Joshi (1981) notes how in a tribal development block in Andhra Pradesh, forest guards take a share of the value of minor forest produce collected by the tribals, and sometimes make the people work without wages. Again from her survey in Kushmi block in Madhya Pradesh she finds that guards there collect regular tributes from the tribals; non-givers are implicated in legal cases with the connivance of the local police. Chand and Bezboruah (1980) from their survey in Betul district (Madhya Pradesh) note how tribals are frequently asked to work for the forestry officials without wages, in return for legalising their right to the land on which they have in fact worked for years. Swaminathan (1982a: 342) mentions numerous cases of tribals in Rajasthan being punished by forest guards, at the smallest transgression into the so-called reserved forests: 'On the slighted pretext, suspicion or desire to teach someone a lesson, the forest guards beat up *adivasis* (tribals) mercilessly'. Many tribals are made to pay huge fines (for which no receipt is given) for minor offences.

Similarly, in the Azad Kashmir region in Pakistan, Cernea (1983) notes that there are 50,000 cases of forest offences from the region pending in courts, with one family in every six being implicated in a reported offence. In Bangladesh again, a ten village study reveals how the landless in the villages are harassed and fined by local police and forest officials, the fines usually being used for personal benefit (BRAC, 1980: 73-76).*

In many cases initiatives to plant trees taken by the people themselves are found to have been thwarted by forest authorities through lengthy bureaucratic procedures and unhelpful attitudes (for India see UTTAN, 1983, and Swaminathan, 1982b).

Ninan (1983) tries to provide some insight into why guards behave as they do in the Indian Forest Service. She attributes it to low salaries ('a guard who polices several lakh rupees worth of valuable forest is paid only Rs.300 per month') and to hierarchies in the service. Also, she notes that while the duties of the guards have changed today, their conditioning has not: 'What used to be a territorial responsibility concerned more with wood than trees, is now far more complex. Where they were asked to grow commercial timber they are now being asked to produce firewood, provide employment to tribals, grow 'socially useful' trees and what's more, actively persuade rural folk to raise plantations. . . . Persuasion does not come easily to a force used to keeping people out of forests'. Ninan emphasizes the need to reduce inequalities in the service and to crack down on the culture of blatant corruption. How this can be done in a structure built on hierarchies and privilege is a moot question.

---

*It is interesting to note that historically, exploitation and antagonism have often been the basic features of the relationship between forest officials and the local people, even in parts of Europe. In England, for instance, the preamble to the Exmoor Forest Ordinance of 1306 makes fascinating and revealing reading:

'We have indeed learned from the information of our faithful and the frequent cries of the oppressed . . . that the people of the same realm are by the officers of our forests miserably oppressed, impoverished and troubled with divers wrongs, being everywhere molested. . .' (quoted in Burton, 1969: 25).

Also, while practices such as the extraction of fines for minor transgressions mainly involve the forest guards, other practices of corruption are not limited to the guards. For example, in the earlier-mentioned study by BRAC, in Bangladesh, it was noted that even while the landless were being harassed and fined for taking minor forest produce, the trees on government land were being cut illegally at a rapid rate by locally powerful timber merchants, with the unofficial cooperation of the forest officers. What the merchants had done was to buy private plots, situated inside the government forests, from the local tribals at very cheap rates. They had then obtained permits to cut down trees from the private land. This land, however, had few if any trees, and the ones they actually cut were growing on the government land adjacent to their plots, and which they then carried into their own plots. It was noted that 90 per cent of all the trees cut in this area were from government land. The merchants ensured the cooperation of the local officers through bribes. Thus '...a small number of rich and powerful men were able to make large profits from public resources. At the same time, these men were also able to use their government connections, their economic base, and threats and actual use of force to obtain control over the land and the actions of others' (BRAC, 1980: 2).

This is by no means a stray example. Nath (1968) provides a similar case-study relating to government forests in north India, where too the local timber merchants were able to gain access to the land of local tribals, and were making large profits by cutting the teak trees both on this land and on the adjacent forest land belonging to the government. In this study, though, the practice was cut short through the intervention of the young civil servant in charge of the area who also managed to make the tribals conscious of their rights and of the fact that they were being exploited by the merchants, by being paid marginal sums for highly valuable timber. The study of course points to the personal integrity of the civil servant, but more interestingly the description of the ingenuity that needed to be exercised to confront and tackle the merchants serves to illustrate how complex the mechanisms are through which power is exercised by a few at the local level, and how difficult it is to deal with it.

Swaminathan (1982a: 344) raises some pertinent questions in this context: 'For whom are these forests reserved? Protected for whom? Protected from whom?'

All said, both the protection of existing forests as well as schemes undertaken by the government for planting more trees, far from benefiting the rural poor have in fact, in most cases, become the means of further depriving them of their existing rights. Essentially, these projects undertaken in the name of social forestry provide no guarantee that the benefits will flow to the people. This absence of a surety that the benefits of trees planted will go to those who put in the labour to plant them, also underlies the failure of most of the so-called community forestry schemes, to which I shall now turn.

## 2. Community Forestry Schemes

The assumption in such schemes is that the community will actively participate in tree planting. Yet a consideration of the schemes on the one hand indicates that usually little effort is made by the scheme initiators to actively involve the people in scheme conceptualization and implementation; and, on the other hand, points to the unequal pattern of land ownership and control and the power structures operating in the village, which circumvent voluntary participation by the underprivileged.

### (i) *Inadequate Involvement of the Local People*

The significance of involving the people in scheme conceptualization and implementation is highlighted both by the many failures and by the few stories of success. Consider some examples.

In Upper Volta an FAO village woodlot project in ten villages failed even though it was purportedly designed to involve the local people through the village chiefs. In practice, the involvement was nominal. The government officials informed the village chiefs that their village had been selected for the project, and asked them to select the land, organize the villagers and supervise the tree planting. The villagers were merely told by the chiefs where the trees were to be planted once the project details had been formalized. They were not part of the decision-making and did not believe that any of the benefits would accrue to them. When asked who they thought would benefit from the matured trees, most said the government or the forestry service or the village chiefs or the foreign project designer, rather than themselves (Hoskins, 1979b: 18-19).

Again in upland Philippines, Aguilar (1982) notes, on the basis of detailed case studies of four forestry projects, that the top-down approach was the main constraint to the success of the projects. There was minimal involvement of the people in project formulation or running. As a result (a) the project lacked credibility in their eyes; (b) the choice of trees was wrong—they wanted fruit trees not the *ipil-ipil* seedlings that were distributed; (c) they resented not being part of the decision-making; and (d) their immediate needs were for basic subsistence since much of the community was poor. These needs were not met by the project.

In the case of one of these four projects, in the Buhi lake area, Aguilar notes that the project's potential beneficiaries (many of whom participated with the hope of being employed as labourers) took the initiative of forming an Association to express their collective views, but found that the project staff did not give the Association any weightage. For instance, a list of names recommended by the Association suggesting which people should be hired as additional labour on the project, was put aside in favour of persons said to be close to the project staff: 72.5 per cent of the Association members felt that all major decisions were taken solely by the project staff. Hence out of the 130 people who participated only 63 planted trees, firewood development fell short of the target area coverage by 82 per cent, and only 46 per cent of the people claimed to have received any benefit from the project.

That the involvement of the local people in the project conceptualization and implementation is important not only to ensure their interest and commitment but also to ensure that the correct species of trees are planted, is highlighted in other experiences as well. Hoskins (1979a: 3, 4), for instance, describes an FAO study carried out to assess the potential for community forestry in Senegal. The foresters mentioned that they had directed local residents to plant and maintain cashew trees as a fire break around the national forest with the (presumably verbal) promise that they could harvest the nuts once the trees matured. Yet the residents were found to be deliberately destroying the trees so that they would have no responsibility for maintaining them. From this experience the foresters surmised that forestry for local community development* does not

---

*Forestry for Local Community Development (FLCD) relates to a new approach to forestry schemes emphasized by the FAO in recent years: 'This type of forestry should encompass any situation which ultimately involves local people in a forestry activity for the direct benefit of those people' (FAO, 1978b).

*Fuelwood and fodder*

*Cane for baskets*

*Bamboo and straw for huts*

*Medicinal herbs*

*Social forestry must allow for the multiple needs of the people*

work. Yet Hoskins notes that the same Senegalese foresters reported being unable to keep up with the growing demand of the residents to purchase fruit tree seedlings from the forest service nurseries. These trees, which provided the residents with fruit and shade, had a 100 per cent survival rate. Clearly *this* was the real success story of the project.[7]

However, a number of West African woodlot projects are noted to have failed even after they were planned with the local village men who willingly planted the trees. Hoskins (1979a) attributes this to the male bias in extension, and the non-involvement of the women in the projects which were located in regions where women traditionally do the tree maintenance tasks. Since they were not involved in the project the trees died due to lack of care. Again, in Sierra Leone, a tree crop which produces a fruit that women traditionally process was planned and planted by men. Since the women were fully occupied in that season much of the fruit spoiled (Hoskins, 1983: 9).

The importance of involving women in tree-planting projects is in fact highlighted by several other experiences. Of particular note are instances relating to the Himalayan hills of Uttar Pradesh in India. This has been the site of the *Chipko* Movement, one of the most publicized stories of mobilizing people, especially women, for protecting their own environment. In the context of this Movement, it has become apparent that the interests and concerns of the women tend to be much more directly related to ecological preservation than of the men. In one case, for example, in 1980, a government scheme to cut down a large tract of the Dungari-Paitoli oak forest, to establish a potato seed farm and other infrastructure, was strongly and successfully opposed by the local women who resorted to *Chipko* (meaning to cling or embrace) to save the trees. The scheme had been principally supported by the village men (especially those of the *panchayat*) who saw in it the potential for making profits. For the women, however, the cutting of the forest would have meant the destruction of their main source of fuel, fodder and water while cash in the men's hands was seen by them as likely to bring them little benefit. As a result of the women's campaign, forest felling in the area was banned.

Likewise, when social workers in a camp organized in the Chamoli area to involve villagers in tree planting, asked the village assembly what trees should be planted, the men opted for fruit trees and the women for fuel and fodder trees. The women mentioned that the men would sell the fruit and spend the cash on liquor or tobacco, while fuel and fodder trees were essential for family subsistence.

The *Chipko* Movement[8] began as an attempt by the local people to prevent the indiscriminate commercial exploitation of the forests in the region.[9] In this area 95 per cent of the forest land is owned by the government and managed by the forest department which has tended to treat the forests as a source of revenue rather than as a resource for the local people. The specific incident which is said to have sparked off the Movement in 1972-73 is the action by the people in Chamoli district against the allotment of vast tracts of ash forest for felling to the Simon Company, a sports goods manufacturer, while a local labour cooperative—the Dashauli Gram Swaraja Sangh (DGSS)— was refused permission to cut a few trees for making agricultural implements for the community. The villagers, mobilized by the DGSS, resorted to *Chipko* and clung to the trees, challenging the employees of the sports goods company to axe them first. Women were actively involved in this protest. But their special role was further highlighted when in March 1984, twenty-seven of them, on their own, prevented the contractor's employees and forest personnel from felling a large tract of oak forest near Reni village by clinging to the trees throughout the night. Ultimately, the agitation led to the banning of tree felling in an area of 1200 square km.

Since then there is noted to have been a concerted effort in the Movement to end the contractor system of forest exploitation, to demand the banning of green felling and excessive resin tapping, and to agitate for minimum wages for forest labourers. There have been several instances where peaceful protest demonstrations by *Chipko* activists have led to the cancellation of tree auctions (see Dogra, 1984: 49-50). The campaign is now focused both on forest protection and reforestation. In particular, women's role in the Movement has been growing. They have been at the forefront not only in the protest demonstrations but also in keeping vigilance against illegal felling in some areas: instances have been noted of their either having intercepted trucks carrying illegally-felled timber and catching the culprits, or of their informing the forest officials about such activities (CWDS, 1984; Dogra, 1984). In Gopeshwar town the *Mahila Mandal* (women's organization) is seeking to ensure that the forest around the town is protected. Watchwomen who receive a wage in kind keep guard and regulate the extraction of forest produce by people for their daily needs. While twigs can be collected freely, those

*A Chipko meeting in progress: women are active participants in the Movement*

found harming the trees are punished (Jain, 1984). In some villages too the *Mahila Mandals* are noted to be playing an important role in tree protection and new plantings.

The Movement has also brought to the fore the fact that women and men even of the same class of household can have different priorities. The Dungri-Paitoli incident mentioned earlier was one manifestation of this; women opting for fuel and fodder trees with the men opting for fruit trees was another. It is women in this region who in 1979 raised the slogan: 'Planning without fodder, fuel and water is one-eyed planning'. Other issues are also being raised. An anti-alcoholism campaign has been launched. And in some villages women are demanding an equal say with men in village decision-making, especially on forestry issues, and asking: why aren't we members of the village councils? (Jain, 1984). The *Chipko* Movement thus has the potential for growing from an ecology movement to one which calls for an end to exploitation at several levels.[10]

The *Chipko* message has been spread over the years through cross-regional marches by Chipko activists, through the modern media, and through the numerous songs written by a local folk poet and slogans raised by the people. Recently a similar agitation under the banner *Appiko* has been launched in Karnataka State (see Alvares, 1984).

The significance of people's involvement in general and of women's involvement in particular in tree-planting projects is also noted in other success stories. Hoskins (1979a: 41-43), for instance, outlines the main features of a project based at Labgar in Senegal which was noted to have been immensely successful. In it there was close interaction between the project director, the forester, the extension agents (including a woman), and villagers. The village was self-selected by the villagers, indicating their desire to participate. The government and donor agency also agreed to fund other priority needs of the village such as dispensaries, a well with a pump, etc., and the villagers volunteered their labour. The selection of tree species was done by the villagers themselves and there was assurance that the benefits of the trees would flow to the people. The women's involvement and interests were sought to be ensured through the woman extension agent. In other words, here the local residents defined their own goals and the foresters, extension agents, and project director sought merely to facilitate rather than to manage the project.

In China where reforestation schemes have had a considerable degree of success, women have been at the vanguard of the reforestation efforts. In the 8th World Forestry Congress in 1978, Chinese delegates are noted to have said that female tree-planting crews were more successful than male crews: trees planted by the female crews had a 95 per cent seedling survival rate (Williams, 1982).

A related point is the vast knowledge that women often possess of plant species. Fortmann and Rocheleau (n.d.) note how in one experiment in Africa, the local women could identify 20 indigenous species which were highly valued and frequently used by them. The forestry and agricultural personnel, by contrast, were unfamiliar with most of these species, especially with the herbs and shrubs that were not commercial. That there is a store of indigenous knowledge that could be successfully drawn upon by scientists is further indicated by the observation of anthropologists when studying tribal communities. Howes (1979: 3) quotes one such study of a tribe in the Philippines where it was found that as many as 1600 different plant species of which about 400 had not been recorded previously in a systematic way in botanical surveys, could be identified by an average adult. Brokensha and Riley (1980) similarly note, in the context of Kenya, how different groups belonging to the Mbere tribe have acquired highly specialized knowledge of their local vegetation which gives them their food and material resources; herd-boys for example are experts on a range of wild edible fruits on which they often have to subsist. But it is the older women who are noted to be usually the best informed.

All said, it is apparent from the experience of the projects reviewed here, that the closeness with which the community participates in the project can significantly affect its possibility of success.

However, the question which then comes up is: why has there been so much difficulty in bringing about community participation? Can it be located merely, as some scholars do (e.g. Hoskins, 1979a), in the (top-down) approach of the project initiators, or are there other structural underlying factors? In the *Chipko* example, described earlier, the Movement essentially involves hill communities which are not characterized by sharp economic and social differences. Hoskins (1979a, 1979b) too in most of her examples gives the impression that the villagers form a relatively homogeneous group, and the possibilities of divergent interests (based on class or other differences) within the village community are not touched upon.[11] Perhaps this is so in those instances. However, most Third World rural communities, especially those practising settled agriculture, are characterized by high socio-economic inequalities; and the problems of ensuring community participation can usually be traced to the absence of convergent interests, located in the social structure of the community. Also, even to the extent that the problem lies in the *approach*, it is necessary to locate the material basis of the existing ideology and approach. This aspect is considered next.

(ii) *Inequalities in Village Social Structure*

Let us again begin by considering some illustrative examples. Eckholm (1975: 16) relates how a rural reforestation programme in Ethiopia ended in failure because the landless labourers who were in charge of planting the trees, planted them upside down. He says this was because while 'the labourers . . . well knew the difference between roots and branches, they also knew that given the feudal land tenure system in which they were living, most of the benefits of the planting would flow one way or another into the hands of their lords.'

Likewise, consider a tree-planting project in the Azad Kashmir region in Pakistan, described by Cernea (1981). In the project it was assumed that tree plantations would come up on three types of land:

*Shamlat* (community), government and private land. The planting on *Shamlat* land was conceived as a means of ensuring that the main benefits would flow to the small farmers who were in a majority in the community, and who had little access to firewood. It was also expected that *Shamlat* land would promote direct community participation. Planting on the government land was undertaken to demonstrate the benefits of the fast-growing tree species to the farmers, to induce them to plant trees on their private land.

In practice, however, the project was successful mainly in its promotion of private tree planting by the larger farmers. The latter were also willing to invest in the *Shamlat* land but the smaller farmers were unwilling to do so. Cernea describes the underlying reasons for the failure of community forestry. He notes that while in legal terms *Shamlat* land continues to be considered community land, in reality much of it is operated and used as private land. Also, the usufruct benefits from the so-called *Shamlat* land accrue to precisely identifiable individuals—the larger farmers—rather than to the community as a group. The reason is that over the years, much of the land had been informally partitioned for use among some village families whose land enjoined the *Shamlat* land. The plots so partitioned were proportionate to the amount of cultivable land already with the families, and the small farmers thus got little or were excluded. Over time, the rights of use became transferable through inheritance. Some of the land was also privatized as families appropriating it entered it in the revenue records. Hence the tenure status of the tracts of *Shamlat* land (which the forest authorities assumed would provide benefits to the small farmers) had been changed surreptitiously, the lands today being managed on a private basis. Their *de-facto* owners hope to get the lands planted at full government expense with no repayment commitment, and also in the process restrict the current rights to fodder and grass which the small farmers enjoy over this land. Not surprisingly the small farmers are unwilling to put in labour for planting trees on this land. Not only is the prospect of their getting any benefits low but their rights to grazing etc., are also in danger of being lost.

A similar pattern is noted in the Rajasthan and Madhya Pradesh States of India. Jodha (1983) on the basis of a survey of four villages in these two States, notes how in recent years there has been a gradual appropriation of common property resources by large farmers. A good deal of this is attributed by him to the way land reform and other welfare schemes were undertaken: under the schemes considerable areas of common property resources were demarcated for distribution to the landless; but in practice much of the land so privatized went to the larger farmers. In the case of Rajasthan, for example, 59-62 per cent of the land in the villages Jodha surveyed went to farmers who already possessed 10-15 ha. The landless received 11-13 per cent of the land—much of it of poor quality.

This is in fact a familiar story. In most village communities, attempts to use so-called village community lands for purposes of communal benefit are strongly resisted by village factions which have *de-facto* (even if not legal) control over the land. This crucially affects village self-help forestry schemes.

In Gujarat in the earlier-mentioned World Bank-aided forestry project started in 1980, the village self-help woodlot component has largely been a failure. A mid-term appraisal report locates the lack of success to the non-homogeneous nature of the village community, the mistrust in the system regarding its ability to ensure equitable distribution of woodlot output, disputes among farmers on the availability of common land for establishing village woodlots, and so on (see World Bank, 1983a).

In this context, it is noteworthy that the two best-known examples of social forestry that may be termed successful on a country-wide basis, and not just in isolated schemes, are South Korea and China, in both of which the schemes were introduced within relatively egalitarian agrarian structures brought about by radical agrarian reforms undertaken in the 1950s. Of course the ability to implement the reforms or to mobilize the people for community schemes cannot be separated from the political systems of the two countries, nor from their cultural backgrounds.

South Korea in the late 1960s is noted to have presented a grim picture—even the litter from the hillsides was being raked due to a shortage of fuelwood (Eckholm, 1976, 1979). Although the bulk—73 per cent—of the forest land in the country is privately owned, communal access to forests (private and public) has been the traditional practice, and no stringent protective measures were taken by the owners as the land got steadily depleted of its resources (FAO, 1982a). Since 1973, however, the landscape is said to have been transformed—a result of the reforestation programme begun as a part of the *Saemaul Undong* or New Community Movement for Rural Development launched by the government around 1971.[12] The thrust of this Movement has been towards comprehensive rural development through self-help community participation. The emphasis has been on the selection and communal implementation of a variety of projects decided upon by the villagers themselves and agreed to at the village assembly. To promote village participation specifically in reforestation, a concerted effort has been made to establish Village Forestry Associations (VFAs) in every village (while earlier only a few existed), and to ensure that each household head joins the Association. An elected male leader is usually responsible for the overall coordination of the VFAs activities and a woman leader for directing and coordinating female participation in village activities. The VFAs are in turn linked to country and national level forestry bodies which provide strong financial and technical support.

Most of the land that has been chosen for tree planting is privately owned. The owners could opt either for undertaking planting on their own, or in cooperation with the VFAs under varying management and proceed-sharing arrangements. In most cases, the VFAs are today noted to be managing the reforestation efforts completely. Through the VFAs, the whole community is said to jointly decide on tree species, and to communally plant, tend and harvest the wood-lots without pay. The harvested wood is then distributed among the households and the proceeds from the surplus become part of a cooperative fund for further community projects. The VFAs also sometimes participate in business enterprises for the processing of forest by-products.

Eckholm (1979) notes that by the end of 1977, some 643,000 ha of village woodlots, mainly to serve fuelwood needs, had been established in this way. For many households this enabled a switch from purchased coal to fuelwood, effectively increasing income by 15 per cent. The dependency on straw for fuel has also declined, and the country is now noted to be self-sufficient in fuelwood, and moving towards an emphasis on the promotion of wood for commercial purposes (FAO, 1982a).

Why has cooperation succeeded in South Korea? Brandt and Lee (BL: 1980) in their detailed analysis of the functioning of the *Saemaul Undong* in some sample villages, provide interesting insights into this, as does the FAO (1982a) report. From these it is seen that among the noteworthy features of the South Korean success have been the relatively egalitarian distribution of land and wealth, the effective village leadership, a culturally conducive environment, and the noted tackling of reforestation within a comprehensive rural development programme.

In 1971 when the Movement was launched the farm size distribution was largely equitable and most of the farm holdings were extremely small—65.3 per cent being less than 1 ha in size and 99.5 per cent being less than 3 ha in size (and accounting for 93.2 per cent of total farm area).[13] The small size of holdings meant that eacn farmer could not individually and independently conduct all farm activities; hence there was a strong incentive to cooperate. Overall equality meant that such cooperation could take place in practice. The formation of agricultural cooperatives in fact much preceded the initiation of the *Saemaul Undong*.

It is of note though that even within the overall egalitarian land distribution pattern, in those villages where the agrarian structure is especially unequal and where there remains a large proportion of poor and a few well-off farmers, widespread participation has been difficult to organize. The poorer farmers and landless are reluctant to contribute free labour for communal projects since they feel the benefits will go to the large landowners (BL, 1980: 67, 68, 94, 136).

An effective village leadership has also been instrumental in the scheme's success. But here again the effectiveness of leadership is found to be higher where the village social (especially kinship) structure is relatively homogeneous. Where there are sharp structural (usually kinship) divisions within the village it has complicated the problems of leadership and made sustained, concerted action more difficult to achieve (BL, 1980: 135).

It is noteworthy that, once more, the absence or presence of conflicting interests based on class and social groupings—in other words the issue of social structure—emerges as being important in determining the success of communal schemes.

Of course the specific cultural aspects are also important in this success story. Brandt and Lee note: '*The Saemaul Undong* has helped reimpose recognized standards, some of which, like diligence, cooperation and discipline, reinforce patterns of behaviour traditionally sanctioned by Confucian ethics' (BL, 1980: 128). The sense of mutual obligation and solidarity especially within kinship networks (for example, cooperative labour among kin groups is said to be a well-established custom) also forms part of the tradition in South Korean society (BL, 1980: 127).

Most scholars (BL, 1980; Kihl, 1979; FAO, 1982a) agree that while the initial project work may have been done under bureaucratic pressure, over time this has had a tendency to become more voluntary. Kihl (1979: 153) writes: 'What seems astonishing . . . is that the *Saemaul* Movement has changed its character over a period of time—from the coercive mobilization campaign in the early years to a more persuasive, voluntary effort of self-help in subsequent years. In spite of initial setbacks and resistance, the policy goals of the *Saemaul* Movement have come to be accepted by most of the rural population in Korea'.

In China again, community forestry has been described as a success. By one estimate quoted by Eckholm (1979:40), about 30 to 60 million hectares of new forests have been planted over the past two and a half decades, both for environmental improvement and to supply village and industrial wood needs. Eckholm notes that Chinese official figures give an even more optimistic estimate: an increase in area under forest from 5 per cent in 1949 to 12.7 per cent in 1978—in absolute terms an increase of 72 million hectares. As in South Korea, in China too the central government is noted to have made a strong political commitment to community forestry and provided technical and other back-up assistance for this purpose to the local institutions.

Not all the problems have by any means been solved in China. As Richardson, (1966: 60) notes: 'To tend trees for posterity, while freezing in the present from lack of firewood, demands an altruism scarcely expected, even in China', and instances are noted of trees being stolen for fuel before maturity. But these do not constitute the general picture, and the promotion of 'forest conservation and education' and mass mobilization of the people 'to protect forests and trees', is a part of China's forestry programme* (FAO, c. 1979: 25).

Most countries in the Third World (specially in Asia), however, have a highly unequal distribution of the ownership and control of land, which usually constitutes the main source of wealth and represents the principal source of economic and political power, especially in the rural areas of these societies. Land in this context is also one of the principal sources of class conflict. Thus, in the absence

---

*Slogans are a popular form of mass communication on forestry issues in China. Some examples are: 'Cover the country with trees and make it like a garden'; and 'Everything a man can do a woman can do equally well'. This slogan was directed at promoting women's involvement in forestry (FAO, 1982: 20, 25).

*Protection forestry along wheatfields: China*

of redistributive land reform, the bringing about of community involvement in such schemes in most cases is likely to be extremely difficult, if at all feasible. This is borne out not only by the forestry-related examples already given but also by the vast amount of literature on the failure of efforts to promote cooperative ventures in the Indian subcontinent, as well as elsewhere.

In this context, there is in fact a danger of promoting farm forestry, which (as will be seen from the discussion that follows), typically, is not only unlikely to offer a real solution to the woodfuel crisis but can have several undesirable consequences.

### 3. Farm Forestry

The concept of promoting tree planting by individuals on their private land (farm forestry) is in a number of ways similar to promoting a new variety of crop (HYV for example) but it is also different in many ways. It is similar insofar as tree farming has the potential of high profits but involves an initial financial cost, can be constrained by the availability of land, labour and irrigation facilities, and carries with it the risk of financial failure. It is different in that the gestation period for trees extends over several years and not just a crop season; hence the benefits cannot be realized as quickly, and the element of risk is higher. But precisely due to these characteristics of tree farming it is unlikely to be practised by other than the better-off farmers, and further in any large-scale undertaking there will be a tendency to obtain quick returns and grow commercial tree species. This is in fact borne out by available evidence.

To begin with, consider two sample surveys relating to the villages of Rajasthan (an arid zone of India) which have examined the characteristics of households planting trees and those not doing so.

In one survey by Bharara and Sen (1970) which was spread over 10 villages and covered 446 households (of which about 65 per cent owned land and were cultivators and the rest were landless) it was noted that only 8.3 per cent of total households had planted any trees over the three years preceding the survey; of these 75 per cent were land owners, the bulk of whom owned over 20 hectares of land. The number of trees planted tended to be related positively to farm size. The few landless households that planted trees did so on common land or in their home compounds, unlike the landed households which used their own fields. When the households (including those not planting) were asked what their desired place for planting would be, the preferences stated were for field boundaries, near the irrigation well, or in the home compound. Few indicated the communal places—roads, schools, *Panchayat* headquarters or grazing land—as desirable locations.

The bulk of the trees planted (92.5 per cent) were fruit trees and most of the others were planted for providing shade. When all the households were asked what their main reason for planting would be, the provision of shade emerged as the most important factor, followed by wood for agricultural implements and trees for fruit. Only 17.7 per cent of the households felt the need for planting for fuel. This was despite the fact that most of the households met their fuel needs with wood from their own or other people's fields, supplemented by cattle dung.

Those households that did not plant trees, listed their main constraint as being the shortage or non-availability of land. A lack of labour time and water were also important factors. A small percentage indicated that they were not interested in tree planting.

The other survey (Bose and Bharara, 1965) relating to 11 villages and covering 243 households revealed that only 14.4 per cent of the households had planted any trees during the three years preceding the survey. The need for trees was recognised since trees and shrubs were extensively used for agricultural implements, building huts, fuel, fodder and making ropes. As in the first survey, the households indicated that they wanted trees primarily for agricultural implements and secondarily for shade and fruit/food; only 9.5 per cent felt trees were needed for fuel. Most of the needs for fuel were fulfilled through people's own fields, or from common land, and occasionally from other people's fields. Only a few households purchased wood for fuel, though the purchase of wood for making agricultural implements or for housing was common. The main reason given for not planting trees was the absence of irrigation facilities and the fear that a drought would kill them.

The two surveys indicate that one of the main factors constraining private tree planting was lack of access to land, labour and water. Also, wood for fuel was not perceived as an important need. This was in spite of the noted fact that a large number of households had to extensively supplement wood with cattle dung for fuel. A number of reasons could be suggested why in spite of the *revealed* shortage, firewood was not actually perceived as a problem. First, since the area was arid and most farms had no irrigation, the alternative use of cattle dung as manure was probably not very important, since the productivity of manure in the absence of water would be low. Also, the absence of water increased the risk of growing trees, since the chances of the trees dying were higher. Second, the pinch was really felt by the villagers in terms of wood that had to be purchased, that is, wood used for agricultural implements; fuelwood and cattle dung were still 'free' goods. Third, the questions were addressed exclusively to the male heads of households, whereas it is women and children who are usually the wood gatherers. If the women had been asked, a different listing of priorities for planting trees is likely to have emerged.

In more recent years, the much quoted 'success' stories of forestry projects in the Indian context, have in fact related to the planting of trees on private land for commercial use, such as in Gujarat and Uttar

Pradesh (UP). In Gujarat, for example, under the World Bank-aided project mentioned earlier (see World Bank, 1983a), the actual area planted under farm forestry between 1980-81 and 1982-83 has been about 32,000 ha which is about twice the area targeted (while the achievements in village self-help woodlots have fallen short of targets by 57 per cent). The majority of these private plantings have been undertaken by the better-off farmers on irrigated land. The species planted (eucalyptus is the favourite) are fast growing and commercially in great demand, as the prices of poles and pulpwood have escalated in recent years. The actual gains from such planting are estimated to have accrued only to 6 per cent of Gujarat's 2.4 million farm families. While tree planting on government supervised woodlots are also noted to have been successful there is no certainty that this wood will be made available to the people and at a low price.

A similar pattern is found in the World Bank-funded project in UP. Here too, the mid-term appraisal report (see World Bank, 1983b) notes that between 1979-80 and 1982-83, the actual area planted in farm forestry is 3433 per cent of the targeted area (while village self-

G. Foley/Earthscan

*Eucalyptus cultivation on a Gujarat farm: the private profits are high but is this 'social forestry'?*

help woodlots have fallen short by 92 per cent). The most responsive farmers are those with holdings of 2 ha or more and the trees planted are again commercial species for providing building poles.

Both these projects have been termed 'social forestry' projects but in effect have done little to satisfy the purported aims of social forestry, namely the provision of wood etc. for the domestic needs of the rural people. In fact the schemes can have several dangerous social consequences.

First, the shift of large areas of fertile land (irrigated and rainfed) from food to commercial wood production will reduce the availability of food and fibre in the country. These shifts have been observed not only in the UP and Gujarat contexts but in other States of India as well. In the semi-arid Kolar district of Karnataka, for example, Bandyopadhyay (1981) notes that several thousand hectares of agricultural land have been shifted from crops to sericulture and commercial forest plantations. Today, in Kolar district 16,216 ha (of which 13,963 ha is irrigated) is under mulberry, about 20,000 ha under eucalyptus and 6,710 ha under casuarina. The most significant shift has been from *ragi*, the staple crop of the people, to eucalyptus: farm area under *ragi* in this district is observed to have declined dramatically in recent years. Eucalyptus is also the species that is planned to be planted on 45.5 per cent of the 110,000 ha of farm land that is to be afforested under a World Bank-aided social forestry project in the State (Bandyopadhyay, 1981).

Second, the shift from food crops to trees will reduce employment because tree production is less labour intensive. In the proposed World Bank-aided Karnataka project it is estimated that eucalyptus monoculture will lead to a loss of 137.5 million person days of employment (Bandyopadhyay, 1981). Third, the noted shifts are likely to be ecologically destructive where the trees planted are fast-growing commercial species noted to deplete the soil of nutrients and water. Food crops, in many of these areas, are at present grown in mixed or rotation farming (e.g. a *ragi*, pulses and oilseeds rotation) which maintain soil stability. Fourth, farm forestry will have a particularly adverse effect on the poor due to lower employment opportunities, faster-rising food prices, and less access to crop waste for fuel. (Eucalyptus cannot be used for fodder and is unsuitable as a cooking fuel since its burning velocity is too high.) In other words, farm forestry is not only unlikely to solve the problem of woodfuel shortages but will probably accentuate the crisis for the poor.

Yet such schemes are being promoted with short-sighted zeal. The publicity given to the Gujarat project as *the* success story of 'social' forestry in India also reflects a misuse of the term. Take, for instance, the earlier quoted mid-term appraisal report for this project, which despite noting that farm forestry has been essentially commercial forestry in Gujarat, and that village self-help woodlots (which were the main hope for supplying fuelwood) have failed, still concludes: 'It seems possible, that if the present momentum is maintained, it will be possible to resolve the rural fuelwood crisis in Gujarat within a decade' (World Bank, 1983a).

*Mark Edwards/ Earthscan*

*High technology irrigation for a eucalyptus plantation in Niger: who benefits? who pays?*

In many cases the schemes are being defended on grounds of profit maximization. Bapat (1983), for instance, notes that some farmers in Gujarat can earn Rs. 15,000 per acre annually with eucalyptus relative to Rs. 1000 per acre per year with groundnuts, or relative to Rs. 7000 per acre per year obtained by the Punjab farmers with rice and wheat; and concludes that this profit motive must be encouraged still further. The more subtle argument which has been put forward is that after a point, with a glut in eucalyptus, farmers will diversify their plantations and grow other species (e.g. Jha, 1983b; World Bank, 1983a). The weakness in the latter argument is that since it is the profit motive that is sustaining eucalyptus production, it is unlikely that slow-growing trees, which will not yield similar short-term profits, will be planted subsequently. Also, yields (whether of other tree-crops or agricultural crops) on the land left depleted by monoculture eucalyptus plantations are likely to be low unless adequate investment is undertaken to restore soil fertility etc.[14]

The fact is that the current farm forestry policy in India is not the answer to the problem of woodfuel shortages.[15] At the same time, the issue of *social* forestry is intricately linked to the social structure and mechanisms of socio-economic control which determine who benefits how much from the use of scarce resources. It cannot, as Agarwal (1983b) rightly notes, 'be reduced to a simple ritual of planting trees'.

## 4. To Sum Up

It is amply clear that the success of social forestry schemes—viz., the planting of trees to increase available supplies of fuel, fodder, etc. for the local population—is crucial for alleviating the woodfuel crisis. Efforts to increase the efficiency of wood-stoves and to promote improved wood-stoves, as already noted, can make only a small dent. At the same time, as with many other rural development programmes, the issue of increased wood production needs to be structurally linked to that of distribution if the schemes are to fulfil their intended aim.

De-linking the two aspects, as has happened in the thrust towards farm forestry in India, has meant not only that the benefits of increasing production go to only a few, but that the product itself is inappropriate for fulfilling domestic fuelwood needs. Social/community forestry provides the potential for making this link between production and distribution. But the success of social/community forestry schemes, as noted, requires close community involvement and participation, ideally at all levels and at every stage in scheme implementation, from its inception.

Such involvement is found to have been circumvented in most cases by (a) the top-down method of scheme implementation which, in turn, is related to the characteristics of the implementing bureaucracy, such as the method of functioning of the forest department and the attitudes of the forest officials; and (b) the hierarchical socio-economic structures of the communities in which the schemes are located. We have seen that the causes for project failure lie not in the antagonism between people and trees, but in the antagonism and differential interests between people and people: between the forestry officials and tribal communities; between extension workers and the village poor; between different class and kinship groups in the village, and so on. Successful projects have emerged precisely where the material basis for such antagonism (especially in terms of land distribution patterns) has either been eroded, say through a radical agrarian reform programme, or has not existed historically to the same extent (say among more egalitarian hill communities).

The role of such structural factors in scheme success or failure is beginning to be recognized in the evaluation reports of scheme-promoting agencies, but the solutions offered rarely go beyond the conventional. For example, an FAO (1977: 14) report notes that the 'challenge to foresters in programmes for rural communities is evidently an institutional rather than a technical one. The overriding problems are usually not problems of species or techniques, but of relating to the community situation and an ability to motivate people and encourage the necessary political will. . . . To respond to these challenges may require radical changes in the training that foresters traditionally undergo'.

FAO's recognition that the problem is not one of *techniques* but of community participation is clearly a step forward, but its emphasis on training is inadequate. In most cases, for instance, it is doubtful that training alone can alter the deep-rooted biases that the foresters display. Hoskins (1979b: 14) notes, for example, that: 'Foresters in Senegal say repeatedly that women *cannot* be involved in projects as they *do not* and *cannot* plant trees, when Senegalese women have traditionally raised crops as well as planted trees in the courtyards of these foresters' own homes' (my emphasis). Also, there are limits to what even a better trained or committed forester can do when the problem lies in the power groups within the community. Here the issue is not how to activate the people *per se* but how to break the concentration of economic and political power so that true community cooperation is feasible.

As a first step, in large parts of Asia, this may well require re-distributive land reform. It is also essential for most Third World countries to question their existing thrusts in economic development policies and strategies which are marginalizing large sections of the people and destroying the environment. However, such initiatives are unlikely to come from the top in most countries. So far challenges and pressures for radical change have been provided typically by grass-roots movements and organizations of the poor. In this context, ecology movements such as the *Chipko* and *Appiko* in India also have an important role to play, and represent small but significant steps forward in the right direction.

## Notes

1. The people of the Himalayan region of India have identified their needs as the five 'Fs': food, fodder, fuel, fertilizer and fibre (Dogra, 1984).

2. Adeyoju (1976) provides a useful summary of different types of tropical forest tenure, especially in the context of East Africa, in terms of forest ownership and control patterns. He also discusses the possible implications of different tenure systems for people's traditional rights and for tenurial stability.

3. Around 1978-79 this led to several militant clashes between the tribals and the police.

4. Another area of recent public debate has been the 1980 draft bill of the Indian Forest Act which, among other things, would have empowered the forest department to declare any area (not just natural forests) as forest land, made entry into an area declared 'reserve forest' without a permit a crime, and given wide-ranging powers to the police and forest officials to punish offenders. The bill was strongly and widely criticized especially by those working with tribal and forest communities, who emphasized that it would destroy the tribal mode of life and economy and convert forests basically into raw material producers for industry. As a result of the public pressure the bill has been shelved (see Chowdhury, 1982a, and Kulkarni, 1983).

5. The song has been composed by folk poet Ghan Shyam Shailani.

6. Several State-instituted schemes are on-going today which seek to combine tree planting with either crops (agri-silviculture) as in Kenya and Thailand, or with pasture (silvipasture) as in the Solo river basin in Indonesia, and in Nepal and the Sahel (FAO, 1977).

   However, the example from Java discussed above illustrates that practices such as agri-silviculture or silvipasture which provide a 'technical' solution for the problem stemming from competition between alternative uses for land (e.g. for cropping, grazing and forestry), may not succeed unless the socio-economic context in which the schemes are promoted is also appropriate.

7. For some additional examples (from India) of success stories where people have been purposively involved in scheme implementation see Sarin (1980), and Chowdhury (1982b, 1983).

8. For details see Dogra (1984), CWDS (1984), Bahuguna (1984), Jain (1984), and Misra and Tripathi (1978).

9. The *Chipko* method of protest by villagers against tree felling is said to have an old history and to have been used for the first time over two centuries ago in the desert area of Rajasthan, and in 1930 in the Himalayan Tehri region as well.

10. Also see Omvedt (1984). In particular, Omvedt points out that movements relating to ecology and environment in rural India are strongly based among tribal and peasant communities, and are people's movements with much radicalizing potential. They also hold a special significance in their raising basic issues relating to the nature of economic development itself and in their underlying demand for the restoration of a balance between society and nature.

11. This is with the possible exception of the Upper Volta example where the village was dominated by the village chiefs.

12. The discussion here summarizes some of the basic features of this programme and is based essentially on FAO (1982a), Eckholm (1979), and Brandt and Lee (1980). These books provide a much more detailed discussion.

13. See FAO (1982a: 9).

14. The issue of farm forestry has been the subject of considerable debate in the Indian newspapers especially in 1983.

15. Both Agarwal (1983b) and Guha (1983) argue that forestry policy in India is currently being dictated largely by the demands of the paper industry. There is also noted to be an acute shortage of pulpwood for the rayon industry and of building poles.

    Guha (1983) further adds that the thrust of international aid-giving agencies is in the same direction. For instance he quotes an USAID report to the Indian government recommending that 'it would be highly advantageous for the Indian economy to replace a significant percentage of the mixed tropical hardwood species with man-made forests of desirable species such as eucalyptus, tropical pines and teaks' (USAID, 1970, quoted in Guha, 1983: 1889).

# CHAPTER VI

# ON EVALUATING WOODFUEL DIFFUSION PROGRAMMES

*They arrived already knowing everything. . . . They come here and look around, but they **see only what is not here**.*

—Villagers to Feuerstein (1979: 2)

*Babuji (Sir), someday you'll understand. It is sometimes better to lie. It stops you from hurting people, does you no harm, and might even help them.*

—A Villager in Punjab
to Mamdani (1972: 19)

How do we evaluate woodfuel diffusion programmes? A consideration of this issue is integral to evolving appropriate directions and means of alleviating the woodfuel energy crisis. In broad terms there are two aspects to such an evaluation. First, we need to evaluate *ex-ante* the relative appropriateness of alternative ways of achieving a set objective (in this case reducing the burden of the woodfuel crisis) in order to make the most appropriate choice. Second, having identified a scheme (or a set of schemes) we need to evaluate *ex-post* (that is, after the scheme has been launched) whether or not it is being implemented effectively, and pinpoint the possible causes for its success or failure.

It is argued here that the effectiveness of either aspect of evaluation will depend on the approach by which the evaluation is done (top-down or participatory), by whom it is done (the local people or outside evaluators), the method by which the information needed for evaluation is obtained (questionnaires, group discussions, etc.), by whom the information is gathered, how it is interpreted, and so on. These aspects are not entirely separable: for example, a participatory approach also implies the involvement of local people in evaluation and in some measure the use of group discussion methods. Let us consider the issues in turn.

## 1. *Ex-Ante* and *Ex-Post* Evaluation

### (i) Ex-Ante *Evaluation: Choosing between Alternative Schemes*

Any attempt at seeking 'solutions' to alleviate the woodfuel problem is likely to involve a set of choices between alternatives by whoever is making policy decisions—say, the government of a Third World country (abstracting for the moment from the question of whether or not particular governments *can* or *will* undertake certain options, given the interest groups they represent). The problem may be that of deciding between the correct 'mix' of woodfuel innovations—for example, whether to devote more funds to the development and promotion of improved wood-stoves or improved charcoal kilns, or to the establishment of fuelwood plantations; or it may mean deciding between woodfuel and other renewable energy sources (say, biogas); or between renewable energy sources (woodfuel or other) and non-renewable energy sources (kerosene etc.); or between providing energy for domestic use or energy for irrigation (which could increase crop production and so the availability of crop wastes for fuel); or between a programme for alleviating the woodfuel problem directly (through woodfuel innovations) and a programme for income generation (which could make an impact on the poverty status of people and hence their ability to spend on fuel), and so on.

These choices acquire complexity not merely because of the range of alternatives which need to be considered but also because the choice between two options implicates larger issues which are not always apparent at first sight. For example, consider a choice between different sources of energy for domestic use, such as firewood, oil and electricity. Investment choices relating to these could also imply a choice between indigenous and imported sources of energy and/or between indigenous and imported technology, and hence have implications for these countries in their degree of dependence on, and vulnerability to, the vagaries of the pricing and supply conditions of such imports in the world market.

Again in planning tree-planting schemes, among other things a choice is involved between alternative uses of land—some of the important alternatives, apart from tree-planting, being crop cultivation, cattle grazing and industrial and urban construction. This would have differential implications for people dependent on particular occupations for their livelihood. Similarly, underlying the relative emphasis given to improved wood-stoves as versus improved charcoal kilns would be an implicit differential weightage given to the users of wood-stoves relative to the producers and users of charcoal. Also involved in the choices would be decisions that are both financial and non-financial in nature; and underlying these decisions can be a complex set of explicit and implicit choices relating to a wide range of connected policy issues, including (at the macro-level) issues such as land reform or other measures that affect the structure of the economy. In other words, every choice made would involve not only technical considerations but also economic and socio-political considerations, and would have a range of possible implications, including the gains made or losses suffered by different sections of the population. The more explicit and detailed the spelling out of these implications is, the more comprehensive can the decision-making process be.

Many of these considerations also arise at a somewhat more micro-level, say in the planning of energy use by and for a village community. For effective and adequate planning in fact all aspects of energy use as well as broader aspects of poverty, development and distribution need to be considered. Energy shortages, as noted, are not suffered in equal measure or for the same uses by all households in a village: landless households may need priority in energy for cooking while landed households may need energy essentially for irrigation or other agricultural uses.[1] Also, as seen in several examples in the last chapter, often for a woodfuel scheme to work, other subsistence requirements of the poorer households may need to be satisfied as well. How the end product of a scheme would be distributed is again an issue that must be an integral part of the pre-scheme evaluation of alternatives.

Unfortunately, most schemes are selected without considering the whole range of options and possibilities and often in isolation of other development projects and schemes. The weakness of a piecemeal approach is often manifest in scheme failure.[2]

Once the choice of an appropriate scheme or set of schemes has been made, there is need to evaluate the schemes launched. The issues arising in this context are discussed below.

## (ii) Ex-Post *Evaluation of Specific Schemes*

A useful distinction in programme evaluation is that between continuous monitoring which essentially implies on-going assessment, possibly of very specific aspects of the programme, and a *longer-term appraisal* which would imply a much more wide-based and comprehensive assessment of all facets of the programme, after it has been on-going for some period of time. The more common approach is to try and gain wisdom after the event, that is to understand say why stoves have not been accepted or a forestry programme has failed. However, continuous monitoring can prove essential for pinpointing weaknesses in the immediate functioning of specific aspects of the programme, in order to correct them where possible. This becomes especially important where the programme involves diffusing a technology which requires adaptation to the user's needs as, for instance, in the case of wood-stoves: here close monitoring would need to be an essential part of the stove adaptation process to resolve problems as they come up (say from a mismatch between stove design and cooking practices or other specificities of user requirements), especially in the initial stages of the project. Such interaction also reduces the distance, in time and space, between technology design and its diffusion.

At the same time, monitoring can be seen as complementing but not substituting for long-term appraisal. Monitoring is essential for identifying and resolving problems located in the specific and somewhat narrow context of the scheme, but long-term appraisal would also be necessary (a) to identify the macro-impact of the scheme (say, the effect of stove-promotion on overall fuel saving in the area, on nutritional patterns, and so on); (b) to bring to bear on the evaluation the experience of schemes undertaken elsewhere; and (c) to relate the experience of the specific scheme to the wider developmental issues and aspects such as the social structure or land use patterns etc. which have evolved through a historical process of change. Such analysis cannot be done on the basis of monitoring alone, which is why the latter can supplement but not substitute for broader-based appraisal.

However, in both the monitoring and the long-term appraisal of rural woodfuel-related programmes, a significant common feature would need to be asking the question: is it reaching and benefiting the poorest and most underprivileged sections (in terms of say class, caste, gender, etc.) of the rural population who are likely to experience the woodfuel crisis most acutely. If the answer is no, it would be necessary to locate the constraining factors: technical, economic, socio-political, etc. And if the answer is yes it is reaching the underprivileged, it would be important to ask what the enabling factors in its success are. An understanding of the reasons for both failures and successes would be important not only for improving the implementation of the particular programme being evaluated but also for planning such programmes elsewhere.

## 2. Who Evaluates and by What Approach?

The quality of an evaluation can depend crucially on who does it and the approach s/he uses. Technically it is possible for evaluation, whether *ex-ante* or *ex-post*, whether in the form of continuous monitoring or long-term appraisal, to be carried out by 'outsiders' (that is, by those who are not the potential beneficiaries), or directly by the people for whom the scheme is meant, with limited outside involvement. In the case of long-term appraisal the outsiders may even be those who have not been associated with the planning or implementation of the scheme at any stage, but who may have specific expertise. Monitoring, however, would require that the outsiders are insiders at least in the sense that they have been directly involved in scheme implementation.

Further, the approach followed in evaluation may either be top-down or participatory. If the potential beneficiaries have been involved in the project as a group and the decision to launch a particular scheme, be it for social forestry or the promotion of stoves, is taken by the group jointly through discussion etc., then implicit in this is a participatory approach to (in this case *ex-ante*) evaluation. On the other hand, if evaluation is done by outsiders, then the approach may be either top-down (with no involvement of the potential beneficiaries) or participatory. However, there are also different degrees to which such an evaluation can be participatory. Participation could range, for instance, from the outside evaluator merely speaking to the potential beneficiaries before arriving at an assessment or deciding which direction the project should take (this latter if the evaluator is also a project implementer), to the outside evaluator merely serving as a facilitator to help the group of potential beneficiaries self-appraise the project jointly as a group.[3]

Whether or not the evaluation procedure is participatory and the degree to which it is participatory, can crucially affect not only the accuracy of the evaluation but also often the success of the scheme itself. Consider first the *ex-ante* evaluation of alternative schemes or of specific aspects of the scheme chosen. In the previous two chapters we noted numerous instances where the priorities of the potential beneficiaries were different from those of the implementing authorities, as in the use of land for cropping or grazing vs. the use of land for forestry, or in the choice of trees, and so on. In the case of projects where the adoption of a technology is a matter of individual choice and involves the use of individual resources (land or labour), a limited *ex-ante* participation of the beneficiaries in terms of consulting them on specific aspects may perhaps suffice. But where the use of communal property such as community land, or a communal contribution of labour and a communal sharing of the products of that land or labour is involved, such a limited participation would be inadequate. Here the group concerned would need to be an active participant in the choice between alternatives, as indicated by the experience of many of the forestry programmes reviewed earlier.

Next consider the issue of *ex-post* evaluation. Here again the success of the scheme can depend closely on how the evaluation is attempted. For instance, as noted in the diffusion of technologies such as stoves, effective monitoring will require a close interaction of any evaluator with the user, to understand why, for instance, the technology is not being used, or is not working as desired. Through such participatory monitoring, appropriate design modifications can be made; it can also serve to acquaint the user with different aspects of repair and maintenance. Participatory monitoring may thus prove to be a crucial component in diffusion. Unfortunately, most projects are monitored or appraised in a largely top-down fashion. (The appraisal 'missions' organized periodically by international funding agencies are commonly of this nature.)

Ideally, monitoring of schemes needing group action or concensus should be done by the beneficiaries themselves as a group, or through a close interaction between the scheme implementers and the beneficiaries. However, in long-term appraisal it may be advantageous to supplement self-appraisal by the group with appraisal by someone who is a complete outsider to the project, although not a stranger to its social, economic or cultural context.

One of the significant advantages of group self-appraisal is that the group would have a first-hand familiarity with and knowledge of the social and cultural context of the scheme, a familiarity which most outsiders would need to gain. Also the acceptability of suggestions on changes in programme direction, thrust or content would be ensured if the decision to do so evolves out of group discussion. Additionally, as highlighted in discussions between several grass-roots activists and rural development workers in a recent workshop in India (summarized by Bhasin, 1984), evaluation can be seen as 'reflection on action'. It was also noted at the workshop that such self-appraisal helps to improve the inner working of the group, eases tensions, gives the group greater cohesiveness and helps it evolve a common perspective. It also gives tangible results as action by the group becomes more conscious and focused. To this may be added additional long-term advantages such as the group evolving as a

*South Asian activists appraising their rural programme experiences*

politically conscious one and cooperating on other developmental/economic ventures. At the same time, evaluation by an outsider can be helpful for getting an 'objective' assessment and for bringing a macro-perspective and a wider-ranging experience to bear on the subject. In such evaluation too, participatory techniques such as group discussion can be used effectively. Also, the feasibility of group self-appraisal and group interaction depends on the existence of some degree of initial cohesiveness and homogeneity among group members. Group cooperation for action and evaluation is found to work best where the group is small and relatively homogenous in terms of class, caste, gender, etc.[4]

### 3. Gathering Information Needed for Evaluation

(i) *Nature of the Information to be Gathered*

Whether for making an appropriate choice between different types of schemes and programmes or evaluating the working of specific schemes, one of the crucial inputs is the availability of accurate and wide-ranging information on a variety of aspects. The nature of information would usually contain several components such as (a) technical (including physical) information say regarding the amount of different types of fuel, including woodfuel, consumed by different households for various purposes in different seasons of the year, or information on energy consumption for all purposes in the household or village, or on the time taken to gather fuel; (b) socio-economic information such as on property, land or tree ownership patterns in the village, or on the nature of interpersonal relationships in the village (based on class, caste, political or kinship affiliation, etc.), or on decision-making patterns within the household; (c) cultural information such as on cooking practices, religious beliefs; and so on.

Such information would need to be collected both *ex-ante* and *ex-post*. A base line survey prior to the initiation of any scheme would be needed for the selection of appropriate schemes, for working out details of the schemes selected, and for providing a point of reference for any *ex-post* evaluation of the effects of the scheme launched. A re-survey on selected aspects would help to identify the impact of the project.

(ii) *Technique Used in Data Gathering*

The appropriate method by which such information is obtained will depend, among other things, on the nature of the information needed. Consider first the question of gathering technical information, say relating to firewood use in the home. Such information can be obtained by asking the respondent (memory recall) or by actual physical measurement. Here the most accurate information would be obtained through actual measurement. Basing estimates on the respondents' replies would be unsatisfactory on several counts.[5] To begin with, typically, the information on firewood use supplied by the household would not be in standardized weight units but in non-standard units such as bundles of various sizes, head and shoulder loads, etc. The overall weight of firewood in these bundles or loads could vary between households, and even within the same household, depending among other things on the carrying capacity of the person(s) collecting the firewood. The actual daily consumption of firewood in a household could also vary from day to day—a variation which would not usually be revealed unless actually measured.[6] Again to the extent that firewood is gathered in the form of small twigs, etc. it can only be taken into account adequately by physical measurement. Likewise, where fuels other than firewood are also used in the household they would need to be reduced to a standard unit to assess their proportionate use relative to firewood, and to assess the household's total energy consumption: here again the respondent's estimates would not be adequate. By Arnold's (1980) assessment the margin of error in the respondents' estimates of weekly fuel consumption, if properly coverted into weight, would be within ±20 per cent of the actual average; but he notes that the coversion of non-standard units into weight units presents serious problems.

Unfortunately, few energy demand studies are based on an actual measurement of fuel used (see Desai's 1980 comments on some Indian studies). Among these few are Arnold's (1980) study in Indonesia, Fleuret and Fleuret's (1978) in Tanzania, Bialy's (1979a, 1979b) in Sri Lanka, ASTRA's (1981) in Karnataka (India) and Briscoe's (1979) in Bangladesh.[7]

Of course the physical measurement of household fuel consumption is inevitably time consuming and would be difficult to undertake for large samples or over extended periods of time. In such cases the recall method may need to be resorted to. The same would hold true for the collection of data relating to time taken in particular tasks, such as in gathering firewood or in cooking, since direct observation, say by following people around throughout the day, would considerably limit sample coverage and possibly also affect behaviour patterns. However, in the use of the recall method, the choice of the recall period can make a significant difference to the accuracy of the information obtained.

Bajracharya (1979), for example, compares data relating to the annual fuelwood consumption of a sample of 8 households in a village in Nepal, as obtained from a one-time annual recall and that obtained through weekly records kept by the same households and cumulated for 52 weeks. He finds a consistent underestimation of fuelwood use by the annual recall method in 6 out of the 8 households. Similarly, White (1984) compares data gathered through three different recall periods and relating to the average annual working hours of 60 sample households in rural west Java. He finds that with a 12-month recall relative to a 24-hour recall, over 40 per cent of the average annual working time per household was missed out; and in a 30-day recall relative to the 24-hour recall, one-third of the working hours had been missed out. In general, both Bajracharya's and White's analysis indicate that the shorter the recall period the more reliable is the data likely to be; and where seasonal variations are to be recorded, frequent interviews throughout the year would be desirable.

The gathering of data by season is in fact especially important in relation to firewood consumption since the availability of firewood varies a good deal by season, and in many regions the experience of shortages is noted to be especially acute during the monsoon (e.g. see Briscoe, 1979). According to Chambers (1983: 20) there tends to be a dry season bias in the collection of data relating to rural areas. As he puts it: '... rains don't do for rural study—suits get wet and shoes get muddy.'

In gathering data on aspects such as the consumption of firewood or time allocation, there is also a case for involving the local people either as a team working along with the evaluator or even in collecting specific aspects of information on their own. For instance, in both Islam's (1980) and Bajracharya's (1981) studies, local educated youths helped collect quite accurate data on firewood use on a long-term basis over different seasons. The participation of local people in data gathering would also help increase their interest and involvement in the project.

In the collection of information on socio-economic variables, the issue of how it is collected is again crucial. The most commonly-used method of eliciting information in rural energy surveys has been through the questionnaire, especially of the close-ended type. However, as Barnett (1982) and others have pointed out, such an approach could well impose preconceived notions and ideas on the situation observed, restricting the scope of the evaluation. Also it can miss out on a range of complex qualitative information. Thus, a questionnaire can, in this context, supplement but not substitute for other methods of data collection.

Among the more imaginative and participative techniques used by researchers in specific cultural contexts has been that of adapting the format of traditional village games to generate information and stimulate discussion among the villagers. In Nigeria, for example, the game of *Ayo* played by the Yoruba tribe has been adapted.[8] In Ecuador a game called *Hacienda* has been developed which simulates the setting of an Andean village and deals with issues of land ownership, improved farming, and the role of various local authorities.[9] Such games or other methods which relate to the villager's cultural experience could supplement the more conventional methods such as questionnaires (also see Richards, 1978). This could be helpful, for example, in eliciting more frank answers from the economically and socially underprivileged, regarding their relationships with the large landowners, especially in cases where the former are tied in economic or extra-economic coercive relationships with the latter, and would therefore hesitate to speak openly against the landlords.

The games also serve to raise the political consciousness of the players. For example, in the *Hacienda* game mentioned above, different players, to begin with, command different shares of wealth and advantage. The roll of the dice is seen to reflect 'the accident of birth' by which one person is a hacienda owner and another a labourer. The process of playing can lead to an objectification of the concept of 'fate' and a discussion on *why* one person owns land and another does not (see especially, Smith c1973: 14-15).[10]

Participant observation is also important for gaining greater familiarity with the way in which the local community classifies and categorizes information, which may be quite different from the way a researcher unfamiliar with the area may do so. Richards (1978), for example, describes how in parts of southern Nigeria, grasshoppers which were classified as agricultural *pests* by scientists were found to be an important part of the diet of many people. Categories imposed from outside could thus miss out on crucial information.[11] An analogous example in the context of wood-stoves would be the usefulness or otherwise of smoke: it is often seen as an irritant by technicians seeking to improve stove design while, in practice, in many village households, smoke is useful in helping to keep thatched roofs dry and in keeping out snakes and insects from the huts, etc.

Whether initiated through games or through informal meetings with the potential beneficiaries of the scheme, group discussion has been found to be an extremely effective method for obtaining various types of socio-economic and cultural information and promoting participatory evaluation. The Banki drinking water project in UP (India), which was described in detail in Chapter IV, is illustrative. Here group discussion involving the villagers helped both to understand why an earlier scheme (which had used a top-down approach) had failed, and to promote the acceptance and direct involvement of the villagers in the new scheme.

### (iii) *Information Gathering and People: Interrelationships*

Basically, information on a project affecting *people* would need to be gathered from the affected (or likely to be affected) people, and the scope and quality of information obtained can depend crucially on the reaction of these people to the evaluator. As one researcher with field experience in the villages of Pakistan Punjab points out: 'The very term 'data collection' implies that researchers conceive of 'data' as if it were an object having a *concrete* existence and the job of the field workers is to go and 'collect' it. Such a conception ignores the simple truth that socio-economic data ... emerge out of a relationship, not between the field worker and an object, but ... between the field worker and another human being' (Hussain, 1980: 281). For an outside investigator, establishing a rapport with the villagers is therefore essential, and where the desire is to understand the situation from the point of view of the economically and socially underprivileged, it requires special effort and care to gain their confidence.

The way in which a person is received in the village would be affected by (a) the person's objective situation—for example, by the researcher's gender and whether s/he is an 'outsider' or an 'insider'; and (b) her/his attitudes towards and sensitivity to the villagers. Consider how.

The evaluator's gender becomes important in terms of the person's access to people of the two sexes. A man, for example, would be at a disadvantage when seeking information from women users of stoves; in some countries where female seclusion is practised, a male evaluator may have virtually no access to local women.[12] In a different context the reverse could be true—women evaluators could experience much greater difficulty than male evaluators in establishing rapport with men in a traditional village community.

The 'outsider-insider' issue similarly concerns the evaluator's access to local respondents and may be seen to have two dimensions. In one sense, as already noted, it relates to the person's formal affiliations with the local community or with the project. In another sense it relates to the person's familiarity with the local culture, language, politics and so on. While it is obviously desirable that the evaluator be an insider in the latter sense, it would not necessarily be an advantage if s/he is additionally an insider in the former sense, since *then* the evaluator might also be seen as having vested interests either *vis-a-vis* the project or *vis-a-vis* some local village faction.

Even as an outsider in the latter sense, however, the way in which the person is received by different factions or classes within the village would be influenced by the class bias the person appears to have. Here details such as whether the researcher approaches the village in a jeep, on a bus or on foot, whether s/he makes the initial contact with or resides with the large landlord or the poor peasant,[13] whether s/he wears town clothes or the traditional village dress, can all make a significant difference.

However, for the researcher to be able to establish a rapport with the villagers, whether through games or otherwise, would require her/his approaching them as people from whom s/he can and wants to learn, rather than as someone who already has the answers. Establishing a dialogue with them would need both sensitivity and sensibility. It is these factors which would reflect what could be termed the evaluator's 'political consciousness'. It is this which would determine whether the evaluator even bothers to speak to the people affected, or likely to be affected by the innovation—to speak, for example, to the actual users of wood-stoves or to the participants of forestry programmes for finding out what *they* perceive as immediate problems.

The answers obtained can be quite different depending on who is spoken to. In a forestry project in Africa for example (see Hoskins, 1979b), those involved in the project, namely the international 'donors', the host country technicians, and the local participants were asked what they thought were the causes of its failure: each saw the problem as lying in the actions of the other groups. The donors blamed the villagers for not participating and the foresters for promoting inappropriate technical packages and not ensuring village participation. The foresters similarly blamed the villages for non-cooperation, and additionally the donors for not providing the necessary logistic and materials support. In other words, both the donors and the foresters located the problem on the one hand in technical bottlenecks, and on the other in the villagers' attitudes. Both assumed in effect that the programme was advantageous for the villagers and that their lack of cooperation was, therefore, due to irrational attitudes rather than legitimate causes. In fact, as the villagers pointed out, their non-cooperation stemmed from the fact that they felt the programme did not take their real needs into account, or use their expertise. Further, they were given no assurance that they would receive benefits from the project.

Similarly, whether the evaluator speaks to the women or the men can be of considerable importance, since (as noted) the perceptions, concerns and priorities of the two can differ.

An interesting example of how many of these factors can affect evaluation is provided in the context of an action-cum-research family planning programme launched by the Harvard School of Public Health, in the villages of the Indian Punjab, and known as the Khanna Study (Wyon and Gordan, 1971). In this, two evaluators (the Study directors and an independent researcher) arrived at quite different evaluations of the programme. Even though this example does not relate to woodfuel innovations *per se*, it is considered here in some detail since it illustrates so well some of the points made above.

Under the programme, detailed information was recorded on the lives and habits of the villagers, their family structures, and their economic situations. Regular monthly visits were also paid by field workers to couples of child-bearing age to provide advice, distribute contraceptives and record their use. A follow-up evaluation survey carried out by the Study directors revealed that as an attempt to diffuse contraceptive technology the programme had failed. The failure was attributed by the Study directors to the high infant mortality rates which they said caused parents to hesitate before drastically curtailing births, and to the inability of the people to assess the true impact of their rapidly growing numbers on their personal welfare (Wyon and Gordon, 1971: 290).

A year after the Harvard specialists had gone, another evaluator, Mamdani (1972), undertook a detailed survey of one of the villages in the Khanna Study area, and on the basis of extensive interviews with the villagers themselves came up with a different set of answers. He noted that the introduction of HYV wheat raised labour needs on the farm: the larger farmer could hire in the labour or buy labour-saving machinery but the smaller farmers had to depend on family labour. More children meant only a marginal increase in household costs but a significant decrease in farm labour costs.

Much of the information on which Mamdani based his analysis had already been obtained in the Khanna Study. For example, the latter had noted that families sought to have more sons to till the land, or to contribute to household income by taking up non-agricultural employment, or to fight with their votes in village factions etc. But these findings were merely *documented* by the Study directors. Mamdani saw them as an integral part of the villagers' objective reality and hence likely to affect the villagers' decisions. As a result he came up with revealing answers. '

His approach differed in a number of significant ways from that used in the Khanna Study. First of all, he approached the villagers themselves to seek answers. Secondly, the frank replies he received indicate that he attempted to establish a dialogue with them,[14] and not force an opinion. Hence where it took the Study directors two and a half years to learn that many villagers had accepted contraceptives without ever intending to use them, it took Mamdani a few moments to find out why: 'But they were so nice you know. They wanted no money for the tablets. All they wanted was that we accept the tablets. I lost nothing and probably received their prayers. And they, they must have gotten some promotion' (a villager to Mamdani, 1972: 23). This also indicates that Mamdani was recognized by the villagers as being 'outside' the project and hence as having no ulterior motive in approaching them.

This example highlights two especially interesting aspects, one relating to diffusion and the other to evaluation. As an exercise in diffusion it illustrates the importance of considering the potential adopter's economic and social circumstances (as opposed to personality traits). As an exercise in evaluation it clearly shows that contacting the programme participants themselves in order to identify problems, and establishing a rapport with them such that they are prepared to give frank answers, rather than answers which they feel the evaluator expects to hear, are both significant determinants of the quality of information obtained. Being an 'outsider' to the project in the sense of not being an implementer could also be of considerable advantage. At the same time, however, as emphasized earlier, whether or not the evaluator approaches the villagers, how s/he approaches them are dependent on the person's political consciousness.

An additional point of relevance here relates to the period of time that the evaluator/investigator spends with the people. This becomes especially important where information relating to asset ownership or income levels needs to be gathered. Understatements on these aspects by the economically better-off are common, and accurate information on land ownership patterns is especially difficult to obtain where there are land ceilings. It has been noted that the average size of land holding in a village increases with the length of residence of the investigator (!) In other words, the longer the stay the greater the likelihood of getting accurate information on these aspects (see Chambers, 1983: 57). However, the cross-checking of such information from more than one source can also help.

### 4. Interpretation of Information Gathered

The evaluator's perception would also affect the interpretation given by the evaluator to what s/he observes in the field, or what the users/participants report. For instance, one evaluator may locate the difficulties faced in the diffusion of an innovation only in the technical shortcomings. Another may try and look for underlying structural bottlenecks. Again one evaluator may view things only from the men's point of view; another may also take account of any sexual bias against women.

The evaluator's perception could affect too how 'success' and 'failure' itself is defined. A programme of tree planting which raises the absolute quantities of wood available may be seen by some to be a 'success' while others might ask questions relating to the distribution of the wood between users, and if noting that the poorest households continue to face shortages would be more inclined to pronounce the programme as a 'failure'. Questions such as who adopts, that is to what class and social grouping the adopter belongs, how are the costs and benefits of a communal project distributed between people, etc., all become significant in any such assessment.

A useful illustrative example of how divergent the evaluation of the same project can be, depending on what view is taken of specific aspects of it, is provided by the Gujarat social forestry scheme in India which was described in some detail in Chapter V. As noted, those for whom social forestry is essentially a question of planting more trees (irrespective of the type of tree or the distribution of benefits from the scheme) see the project as a success. Others have pointed to the fact that only a minute percentage of households in the State have benefited and that the project has promoted mainly commercial wood species not fuelwood-yielding species. They see the project as an example of 'anti-social forestry'. This example also illustrates how interpretation can diverge from facts as is, for instance, apparent in the conclusion reached by the World Bank appraisal team that this project is well on its way to meeting Gujarat's needs for fuelwood. The Khanna Study example described earlier similarly shows how the same set of facts can be differently interpreted and analysed by two different evaluators. Not infrequently, though, evaluation remains limited to a description of the project without any analysis whatever (for examples see Bhasin, 1984).

The identification of conflicting interest groups within the community where the project is ongoing, and the spelling out of the implications of a project for these groups, itself is not a simple task. As emphasized earlier, to gather information on social structure requires an understanding of how people relate to one another within the village and within the household, and much depends on the political consciousness and sensitivity of the evaluator.[15]

In evaluations carried out by outsiders, the interpretation process can also be sharpened by feeding back, where possible, the information and the results of the analysis to the potential beneficiaries of the project. This could help to enrich and increase the accuracy of the interpretation by bringing to bear on it the experience of the people concerned, as well as to widen the understanding and consciousness of the people regarding their own situation.

## 5. On Social Cost-Benefit Analysis

Often a formalized evaluation method for deciding *ex-ante* between alternative projects is social cost-benefit analysis (SCBA). A brief discussion on this is warranted here since it is a method that is commonly used in economic analysis (including in rural energy studies), often without an explicit recognition of the value judgements it involves. This tends to give it an air of objectivity that it does not necessarily possess. Some of the issues relating to the value judgements underlying SCBA will be discussed here. At the same time, it will be argued that an explicit recognition of its subjective character can in fact make it a useful analytical tool.

In simple terms, the method involves an aggregation of all the financial and non-financial costs and benefits (non-financial ones too are reduced to monetary terms) supposedly relating to all individuals within a society, for each year of the expected life of the project. The differences between the costs and benefits gives the net social returns for each year. On the assumption that net benefits obtained in the future would be less valuable than benefits obtained today, the future stream of benefits are 'discounted' to give their net present value. The rate of discount used is supposed to be one which reflects 'society's ' 'true' preferences for consumption now as compared to consumption in the future. Between two alternatives the one with a higher net present value would warrant support.

Given that many authoritative theoretical writings already exist, which have discussed at length the various problems underlying the concept of social choice in SCBA,[16] no attempt will be made here to deal comprehensively with all the issues. For purposes of illustration, however, some selected aspects will be highlighted.

Very broadly, the major problems concern two (not unrelated) issues: first the impossibility of coming to an unambiguous 'objective' answer as to what 'society' prefers, given that society itself is not a homogeneous entity, and second the difficulty of giving monetary values to the various items which go into the computation of net benefits.

Both these issues involve value judgements on the part of the decision-maker(s). For instance one problem in determining 'society's' preferences lies in the fact that in most situations, with the undertaking of a project (say, the introduction of a technology), some people are likely to gain and others lose. How do we evaluate these gains and losses? Technically, it could be argued, and is argued, that if net gains are positive the project may be supported as the gainers could compensate the losers through lump sum transfers. In practice, however, it is unlikely that all possible losers could be or would be compensated by the gainers. It is sometimes also suggested that a project could be supported as long as hypothetically such a compensation is possible even if no actual compensation is paid. In effect this means setting aside the distributional consequences of the project in the project choice, and supporting it essentially on grounds of economic efficiency. This practice is sometimes justified by arguing either that the distributional effects are not significant or that the desirable transfers could be achieved through other (say fiscal) measures. The difficulty with the first argument is that the distributional consequences even if slight in the case of specific projects could be cumulative over several projects; and as regards the second, there are significant practical difficulties in ensuring that the required transfers take place through taxation etc.[17] Given this, to ignore the distributional consequences of a project itself reflects an implicit support for the resulting pattern of income distribution.[18]

At the same time, in taking the distributional consequences of a project explicitly into account some set of weights will need to be assigned to the gains and losses. If the gain/loss of one unit of real income by the poor is seen as having a higher value than the gain/loss of one unit of real income by the rich, then we will need to assign a higher weight to the former relative to the latter. The question then is: how are these weights to be decided and by whom? In the literature, a variety of ways of arriving at weights (using different assumptions) have been suggested, but the choices again are not free from value judgements. The same holds true in deciding the importance given by 'society' to consuming now relative to consuming in the future, that is in deciding on the social rate of discount.

Such judgements are again involved when we consider the problem of giving monetary values to specific items in the cost-benefit exercise. Market prices do not always serve as the best choice because, to begin with, many items are not exchanged in the market. Further, even for those items that are exchanged, market prices may not reflect the real cost to 'society' due to project externalities, or the existence of various forms of market imperfections, indirect taxes, unemployment, balance of payment problems, etc. Both these problems of valuation are meant to be solved through the use of 'shadow prices' (also called 'accounting prices')—but the fact is that the computation of shadow prices again involves some value judgements.[19]

Basically in a society which is not a homogeneous entity but consists of classes and social groups with conflicting interests, no single set of values can define 'society's' preferences, or 'society's' objectives. Corresponding to each set of interests would be a different set of values and a different set of shadow prices in keeping with the objectives of that social group or class. And if the government represents the interests of certain classes and/or social groups, the government's decisions are likely to reflect the over-all interests of those classes and/or social groups.

The subjectivity involved in SCBA is recognised in much of the theoretical literature, but in practice, in many of the evaluation exercises, the value judgements involved are often not made explicit. However, once it is recognized that ethical judgements are inherent in the evaluation process and these are made explicit, SCBA can become a useful tool.[20] For example it can help to specify systematically, the variables that should be taken into account; reduce the possibility of important considerations/implications (such as externalities, income distributional issues, etc.) of projects being left out; help to explicitly highlight the unknowns and thus provide some indication of the uncertainty attached to specific valuations; and enable the narrowing down of feasible alternatives. Also, it can reveal the implications of different social objectives, and the inconsistencies between the stated objectives of the government (or whoever the decision-making authority is) and the objectives revealed in the choices it actually makes. For example, if SCBA indicates that as between two forestry projects one will benefit mainly the poor and the other mainly big business and the latter is chosen by the government, it would reveal a class bias in the decision made. If similarly biased choices are made over a series of projects it would provide pointers as to the true intentions of the government (as opposed to those projected say in the rhetoric) and the class interests it represents.

Unfortunately, the imaginative use of SCBA and an explicit

recognition of its limitations when it is being applied to specific issues, including those relating to rural energy, is not common. Barnett (1982: 19), on the basis of his review of several energy-related studies where cost-benefit analysis has been used, outlines some of the limitations that have tended to typify the literature, and points out some areas for possible improvement. Basically, as noted in the first section of this chapter, the quality of the evaluation process depends on making sure that a realistic and full range of options are compared, and the possible direct and indirect consequences of each option traced. It is therefore necessary that, among other things, a study does not limit itself to looking at only one technical option; and, where energy hardware is under consideration, that the technical parameters are based on how the technology operates in practice under field conditions. This would help provide more realistic values and increase the accuracy of the exercise.

## 6. In Conclusion

Both *ex-post* and *ex-ante* evaluation of woodfuel-related schemes are an integral part of effective scheme choice and implementation. No single method of evaluation or data gathering can be deemed appropriate under all circumstances, and usually a mix of different types of evaluation exercises (continuous monitoring and long-term appraisal), different techniques of data gathering (questionnaires, group discussions, games) and different analytical tools would be needed to arrive at an adequate assessment. It is nevertheless clear that evaluation involves subjective judgements at many levels; hence, *who* does the evaluation becomes a crucial consideration. It is also apparent that in the context of woodfuel technologies a participatory approach, with the maximum involvement of the local beneficiaries (in data collection, choice of project, monitoring and long-term appraisal), would increase the effectiveness of the evaluation as a tool for improving the performance of the project in question and of subsequent projects.

## Notes

1. Also see Bhatia (1980) on this.
2. The absence of a comprehensive concern was noted as one of the causes of failure

of some forestry projects in the Philippines (Aguilar, 1982), and one of the factors underlying the success of the reforestation programme in South Korea (Brandt and Lee, 1980).

3. Cohen and Uphoff (1980: 225) review a variety of views on participation and note: 'According to most views of participation, that which is initiated from below, voluntary, organised, direct, continuous, broad in scope and empowered would be the 'most' participatory'.

4. See for instance PIDT (1982) on the experience of promoting a participative approach in rural development programmes in Nepal; and Dixon (1978) on women and cooperative ventures in South Asia.

5. On this also see Arnold (1980: 7, 8).

6. In Arnold's (1980) study for an Indonesian sample based on physical measurement, the daily variation in cooking fuels consumed by the households came to ±30 per cent.

7. Briscoe's (1979) study is noteworthy in its level of disaggregation and detail about different aspects of energy use, and in its tracing of the interlinkages between the technical, economic and social factors that affect a household's energy use pattern.

8. The game is usually played with canna lily seeds and a wooden board with holes in it. In one adaptation the holes are taken to represent likely returns to farm investment. During the interview the farmer is asked to imagine, for example, that he has N 100 (units of the local currency) to invest in his farm and to indicate his likely returns over the next five years. He is given five *Ayo* seeds to drop into the holes representing these returns. Next he is given three seeds and asked to show his best, worst, and expected returns during this five year period. This can lead to more general questions about his decision-making and provide an insight into the constraints he faces (Barker, 1979).

9. See Evans and Hoxeng (c 1973: 11); and Smith (c 1973: 14-15).

10. Informal methods of data gathering are also strongly emphasized by Chambers (1983: 64-70) who gives several illustrative examples to show their effectiveness in relation to the questionnaire approach. He also notes that often 'rapid rural appraisal', if imaginatively and sensitively done, can yield important insights.

11. Some social scientists have attempted to develop sophisticated techniques— 'repertory grids' etc.—that would allow the respondents to structure their replies according to their own views of the world rather than accept possibly alien categories imposed by the more conventional questionnaire (see for example Barker, Oguntoginboj and Richards, 1977; and Barker and Richards, 1978).

12. See, for example, Bertocci (1975); Abdullah and Zeidenstein (1978); and Islam (1982) for several articles on this aspect.

13. Also see Blanchet (1982) on this aspect.

14. The attitude of many of the Khanna Study field staff towards the villagers, is apparent from their reply to Mamdani's query about why they thought the villagers were not practising birth control. Nine out of the 11 interviewed by Mamdani (from a total of 20) said it was because the villagers were either 'illiterate' or 'prejudiced' or that they needed some 'basic education' or 'demographic education'. In other words, all, in one way or another, considered the villagers to be ignorant, demonstrating in this case not their *cultural* bias (they were all from the Punjab) but their *class* bias (most were urban, middle

class)—see Mamdani (1972: 47-48).

15. In most cases, outside evaluators tend to belong to the relatively privileged sections of society—the educated elite, the upper echelons of the bureaucracy, the members of international organisations etc.; they are also predominantly males. Hence in their evaluation much depends on their ability to transcend the consciousness and interests of their class and gender. The rarity of such evaluators would be no surprise.

16. On the theoretical aspects see, for example, Arrow (1963); Ball (1979); BOUIES (1972); Layard (1972); Sen (1970); Stewart (1975); Dasgupta and Pearce (1972).

    Some of the standard texts on the practice of SCBA are Little and Mirrlees (1974); UNIDO (1972); Squire and Van der Tak (1975). Also see Stewart (1975) for some critical comments on the Little and Mirrlees, and UNIDO approaches.

17. See Dasgupta and Pearce (1972: 57 to 61) and Sen (1972: 152, 153).

18. Weisbrod (1972) speaks of the concept of grand-efficiency whereby favourable equity effects are seen as specific benefits of the project. Also see Sen (1972) on the issue of incorporating income distributional considerations explicitly in project design, and on the difficulties of using taxation to achieve the same distributional objectives.

19. Not infrequently, it is suggested that world prices be used as shadow prices (e.g. Little and Mirrlees, 1974). However, this procedure does not present a straightforward way out either. For a critique of the assumptions and value judgement involved in the use of world prices see especially Sen (1972). Also see Dasgupta and Pearce (1972) for a useful discussion on the issue of accounting prices, and Barnett (1982) on some of these aspects as they relate to rural energy technologies. In particular, as Barnett points out at some length, valuation, using any procedure, presents special difficulties in the rural context of Third world economies.

20. For a succinct exposition on this also see Cooper (1979b; especially 195-205).

# CHAPTER VII

# SUMMARY AND CONCLUSIONS

> *Come plant new trees, new forests*
> *Decorate the earth.*
> *Let's relieve the land of the crisis*
> *Come all join together. . . .*
>
> —from a Garhwali song
> sung by *Chipko* activists

In this book, the attempt has been to trace the nature of the woodfuel problem, its manifestations, its likely causes, and the directions in which the means to alleviate it lie. In particular, it has been sought to pinpoint the factors affecting the diffusion of woodfuel innovations—insofar as solutions to the problem are seen to lie in this direction—and the likely pre-conditions for the successful adoption of these innovations.

It has been argued here that the woodfuel problem relates to a set of complex, social, economic and political issues. It stems both from the absolute shortages of wood in many Third World countries and from the maldistribution of available wood supplies between different uses and users. It is inseparable too from the distribution of non-wood-related energy sources between uses and users. Its causes are seen to be linked closely to the way in which land, wood, and energy resources other than wood, have been and continue to be exploited by particular classes and social groups. The possibility of alleviating it is seen to lie, among other things, in the ability of a country to promote a mix of measures (including reforestation and the diffusion of equipment for the more efficient use of wood as fuel) in such a way as to increase the availability of wood-energy to the poorest sections of the population, who are the worst affected by the shortages. However, in many Third World countries, the carrying out of such a programme, in turn, will depend crucially on their being able to overcome many of the biases, especially those predicated on class and gender, which typically plague rural programmes directed to benefit the underprivileged. In this sense, the woodfuel problem emerges as part of the overall problem of poverty, inequality and rural development in Third World countries. This concluding section aims at bringing together these and related issues that emerge from the earlier chapters.

As was noted in the first chapter, at present, in large parts of the Third World, woodfuels (especially firewood) constitute the most important source of inanimate energy even in overall terms, and certainly in terms of their use in the rural areas where much of the Third World population lives. Further, it appears likely that this source will continue to occupy its position of primacy for quite some time to come, given the absence of suitable and readily available alternatives for the use to which woodfuel energy is primarily put— namely, domestic cooking.

At the same time, with the growing depletion of forests in the Third World, the existing supply of wood for fuel has been decreasing rapidly both in global terms, and in terms of its availability in specific areas (particular countries and ecological zones in those countries), and to specific groups in these areas. (Since firewood continues to be non-monetized, most households depend on what they can themselves gather, and levels of consumption often vary significantly by local availability.)

Many of the consequences of woodfuel shortages are already apparent. They are seen in the increase in the time and energy that has now to be spent (mainly by women and children) on firewood collection, the rise in cooking time for women who have to economize on its use, the adverse effect on the nutrition of those who can no longer gather enough for their needs nor afford to purchase alternative fuels, and who are now eating fewer or less nutritious meals, the loss of cattle dung as a manure due to its use as a fuel, and so on. These difficulties have been compounded by the overall ecological ill-effects of deforestation (such as soil erosion, the spread of deserts, the increased frequency of monsoon flooding, the drying up of streams and rivulets, and the growing sedimentation of rivers), by the resultant negative effects on agricultural output, and by the decreased availability for the local communities of forest products other than wood, including fodder. All these problems are likely to become increasingly severe unless counteractive measures are taken.

It is also clear, however, that shortages of woodfuel (which is our primary concern here) are not experienced as a crisis by all. While basically a Third World crisis, it is more acute in some countries of the Third World than in others, and especially so for those who inhabit the arid and semi-arid parts of these countries. Again, both within these ecological zones and elsewhere, the problem is more severely experienced by the households that are economically and socially (by caste etc.) underprivileged. Inequalities in land and tree ownership in particular are found to be related closely to inequalities in the availability and use of woodfuel and other traditional fuels by the household; and access to such fuels from the fields of others often depends on village kinship networks and social affiliations. Among the underprivileged households, women are usually the worst hit since they typically expend the most effort in collecting firewood, while at the same time having an unequal access to family resources, especially food.

In short, the woodfuel problem relates to the shortages of woodfuels suffered on an individual basis by the economically and socially underprivileged in the Third World, and on a community basis by those living in regions where ecological destruction through deforestation is becoming severe. It is a problem both of overall supply and of distribution.

The causes of the crisis, however, cannot simply be traced (as is sometimes done) to the gathering of firewood for domestic purposes. As argued in Chapter I, existing shortages are closely linked to the form and degree of forest use over several decades in the past. Forests have been severely exploited under colonial rule in many countries (as for rail construction, ship-building, etc. in India) and continue to be cleared for purposes such as agriculture, cattle grazing, large irrigation schemes, industrial development, the commercial extraction of timber, and so on. Demographic pressures no doubt reduce the overall availability of land and wood resources in per capita terms, but the uses to which available land and wood are put are, in turn, related closely to State policies and the pattern of ownership and control of land and other assets between different classes and social groups within a country.

It has been noted, for example, that large tracts of forests on private land have been cut and sold as timber, while an increasing

number of landless are forced to survive on diminishing tracts of
'common' forests. Similarly, those who get the permission to cut
public or government forests for timber, or who are able to do so
illegally, usually tend to be persons with the economic and political
power to manipulate the local, administrative systems in Third
World countries. The fuel supply of the landowners can come from
their own land (wood, crop residues, etc.) supplemented by market
purchases. In contrast, the fuel supply of the poor often comes from
twigs and fallen branches foraged from common land, or by courtesy
from other people's land. They are the victims of the crisis, not its
agents.

The problem is also accentuated by the maldistribution of existing
supplies of woodfuels and other energy sources between countries.
One aspect of this relates to the policies of many Third World
countries which export logs or charcoal to the West and to the oil-rich
Middle East, even though this might increase the shortages within
the exporting countries. The other aspect relates to the global
distribution of oil-based fuels, where the advantage lies with the oil
producers and with those who can afford to make their demand
*effective* in monetary terms.

The causes of the woodfuel problem thus lie in an intermesh of
intranational and international factors which determine access to and
forms of utilization of scarce resources. By extension, its alleviation
may be seen to depend on the possibility of bringing about basic
alterations in a complex set of interrelated variables.

Thus far, most attempts to alleviate the crisis have focused on
ensuring greater efficiency in woodfuel use, especially through the
promotion of improved wood-burning stoves, and on increasing the
availability of firewood through tree-planting schemes. Relatively
little attention has been paid to aspects such as a reallocation of wood
between different uses (say, by substituting non-wood materials
for wood in its current uses and thus making more wood available for
fuel), or taking measures to effect a redistribution of wood and non-
wood energy sources between different users (in favour of the poor).
However, what clearly emerges from the analysis (in Chapters III to
V and the Appendix) is that in whatever form or combination of
forms the 'solutions' are conceived, their successful implementation
depends on the degree to which constraints relating to the socio-
economic structures of the communities in which the schemes are
promoted can be overcome.

B.L. Tak

*Containing the desert: planting between sand dunes in India*

During the past few years, for example, across the Third World, over a hundred wood-stove diffusion programmes and a vast number of tree-planting schemes, under the banner of 'social forestry', have been launched. Yet an evaluation of these schemes indicates that few of them have had any degree of success in alleviating the problem of woodfuel shortages faced by the poor. An answer to the causes of their failure has been sought here both by an examination of the characteristic features of woodfuel innovations relative to other rural innovations, and by considering the actual experiences of such diffusion schemes (including the few schemes that may be deemed successful).

To begin with, it is noted that rural innovations can be classified broadly in terms of their technical, economic and social characteristics. The technical characteristics are defined as those relating to the physical features of the innovation; its economic characteristics as those relating to the form—financial and non-financial—of the costs and benefits, and to the gestation period for the realization of the benefits; and its social characteristics in terms of who the adopter is—whether the individual or the community, and if the individual then of what class, gender, etc. A typology of rural innovations on the basis of these characteristics has been drawn up.

Taking account of the experiences of the range of rural innovations that form part of this typology, it has been hypothesized that the factors affecting diffusion, and the ease and difficulty of acceptance of any innovation, would depend a good deal on these characteristics, as would the appropriate approach to diffusion. It is noted, for example, that innovations whose technical appropriateness depends on adaptation to user needs, which require cash expenditure but do not provide cash benefits, which are used by women while men make the decisions on household cash purchases, which are aimed at a large number of individuals or which need communal consensus, and so on, are likely to pose special problems in diffusion. They are unlikely to be accepted through a purely market-oriented approach to diffusion.

Wood-stoves, for example, are usually aimed at persons of all classes, many of whom are very poor and who would have difficulty in finding the cash to buy even a very cheap stove. Further, for many households, the benefits of the stoves would only be non-financial in nature in terms of the saving of women's time and effort in wood gathering and cooking. Such time would not always have an opportunity cost in monetary terms. Also variations in cooking practices and diets would require adaptation of the stoves to the users' needs. Again, in considering tree-planting schemes, insofar as they are on communal land or need communal effort, community consensus and participation is likely to be necessary for making the scheme work. A number of these insights arrived at on an *a priori* basis have helped to supplement (and on some aspects have been reinforced by) the observations based on actual experiences with the diffusion of woodfuel innovations.

The most significant pointer provided by the review of the diffusion literature relating to attempts to promote improved wood-burning stoves, is the importance of adapting the stove to the users' needs and of involving the potential beneficiaries as closely as possible in the designing and building of the stoves. The promotion of stoves through a top-down approach with little attempt to seek the participation of the women who are the main users, is seen as the primary cause of the failure of many stove promotional efforts in Asia, Africa and Latin America. The rare success stories relate precisely to cases where women's involvement in designing and building has been sought by the person(s) promoting the stoves.

User involvement is noted to be necessary for successful diffusion for several reasons:

*Building improved stoves: involving the women is essential*

Phil Hassrick/UNICEF

(i) It helps to ensure that the stove fits user-needs as closely as possible. For example, this enables the identification of the preferred method of cooking (quick cooking vs. slow simmering, or cooking while standing vs. cooking in a sitting position), the preferred time of cooking, the need for space heating, the number of pot holes required, and so on. That is, user-involvement is noted to be necessary for taking account of many of the cultural specificities of user requirements as well as of differences in household needs according to the household's socio-economic status. What is clear is that no *single* stove is likely to be appropriate for all conditions and for all users—a variety of design modifications would be necessary even for households of the same community.

(ii) It generates in the user a more personal interest and therefore a greater acceptability for the project, in addition to enabling the use, and further development through use, of indigenous technical skills and knowledge. High user-satisfaction with the stove and the user's personal involvement in stove building is also noted to have a strong demonstration effect, and to lead to the wider propagation of stoves among the village women through their informal communication channels.

(iii) It encourages the transfer of new technical knowledge from the designer to the user. This emerges as necessary to enable the user to utilise the stove efficiently, to make minor repairs herself, and to adapt the stove to any newly emerging needs. To bring this about, however, it is found important to ensure that the knowledge transferred relates not merely to the stove as a piece of equipment but (where the modifications made so require) to a new process of cooking. Otherwise, the user's attempts to adapt the stove to suit existing cooking practices can thwart the original purpose of the improved design. Further, it is desirable to convey to the user some of the theory behind the improvements—then the user would be in a better position to bring about subsequent improvements independently.

In the context of tree-planting schemes, the top-down approach to scheme implementation is again found to be a significant factor underlying the failure of particular schemes, but here the issue of access to and control over land is noted to impinge equally if not more strongly on programme success or failure. It is observed that tree planting is being promoted primarily in three ways: under direct government management, under community management and under the private management of individual farmers.

A striking feature of many schemes where the government has taken direct charge of tree planting, usually through the forest department, has been the marked hostility typically encountered from the local people. This hostility can be traced to a variety of interrelated factors: the reservation, for such schemes, of land which had earlier been used by the local people for grazing or other purposes, and/or the restriction or termination of the customary rights to forest produce enjoyed by the people, especially by tribal communities living near or in the forest; the planting of commercial tree species which provide neither fuel nor fodder to the people often in the place of existing multi-product species used by them; the lack of effort to involve the local people in scheme planning and implementation even when the scheme is supposed to benefit them eventually; the absence of an adequate assurance that those who contribute labour and effort in such schemes will in fact reap the benefits; the overall antagonistic attitudes of the forest officials towards the local people, and so on. For several of the programmes

*F. Botts/FAO*

*Tree planting in Ethiopia: women play a significant role*

reviewed in Chapter V it is noted that if the participation of the local people in the scheme had been sought, it would have led to scheme success by ensuring a choice of trees in keeping with locally-felt needs and priorities and suited to the ecology of the region, and by bringing about a closer and more personal involvement of the community with the project. Under existing practices, the government schemes for protecting forests or for reforesting land, far from benefiting the poor have, in many instances, further impoverished them, by depriving them of their customary rights and access to forest produce. (Here, the possibility of allotting plots of government-owned degraded forest land or other wasteland to the landless, on an individual or group basis, for planting trees, and giving them usufruct rights to the trees, needs serious consideration.)

The top-down method of scheme implementation is also found to be one of the significant factors underlying the failure of several community forestry schemes, especially in the African context. In some cases, failure can be traced specifically to the absence of attempts to involve the local women, particularly in regions where women traditionally undertake tree maintenance tasks.

But of as much, if not greater, importance in the failure of many community forestry schemes has been the existence of inequalities between households within the community, in their access to and control over land in the village. To begin with such inequalities serve as a barrier to cooperation and participation by the villagers in new tree-planting schemes. Further, in several of the programmes reviewed, the so-called communal land is in fact found to be controlled *de facto* by a few propertied and powerful households in the village. Thus the landless and small landowners have had little interest in planting trees, access to the benefits of which could not be guaranteed under the schemes.

This problem of access to the benefits of tree-planting does not exist in the case of private tree planting by individual farmers. At the same time, farm forestry typically has not provided the answer to the woodfuel problem even in the areas where it has been promoted successfully, as in parts of India. This is because such planting tends to be profit-motivated and is therefore usually confined to commercial tree species; also, it can be undertaken only by the landed. Further, where farm forestry has led to a shift of fertile irrigated land from crops to commercial forestry, it has had a negative effect on employment opportunities, crop output and the access of the poor to crop residues. In the long run, it can even adversely affect the ecology of the region where soil-depleting tree species have been planted.

Problems with the diffusion of improved charcoal kilns have also been considered here, although it is noted that such kilns are unlikely to directly help alleviate the woodfuel crisis of the rural poor since charcoal is primarily a monetized, urban fuel. Indirectly, however, it could help by increasing the income of the producers—most of whom are the rural poor who operate on a small scale, often illegally, and who are usually enmeshed in coercive relationships with middlemen. They generally use the earth kiln method which (although having low conversion efficiency) requires no cash expenditure for producing charcoal. It is noted that in such circumstances the promotion of improved charcoal kilns on an individual basis is clearly not practicable, and the establishment of charcoal producers' cooperatives may be a necessary condition for promoting more efficient methods of production.

But these observations lead to further questions such as: How can such a producers' cooperative be made to work and the grip of the middlemen broken? How can community participation in village forestry schemes be ensured? How can the class and gender biases of village extension workers or foresters be overcome? How can the top-down interaction between scientists/technicians and village extension workers, and between the latter and the technology-users, give way to a two-way mutual interaction? And so on.

These questions would no doubt be answered in different ways by different people. (For instance, there are those who emphasize that the re-training of the village extension workers may be all that is needed to change their attitudes.) And ultimately of course any attempts at answers would need to take account of the country-context within which the scheme is located, and the degree to which specific biases exist and the forms they take.

However, what is clear is that diverse though the issues and problems with diffusion raised above appear to be—underlying them is a clearly discernable common thread—namely, material and ideological inequalities characterizing the socio-economic structures of the communities where the schemes have been undertaken, which lead to conflicting interests within the community.

Cooperation, for instance, is rarely found to succeed among those who are unequal in material terms since it then becomes difficult to ensure an equitable distribution of costs and benefits. In forestry schemes, for example, an initial unequal distribution in the ownership and control of land in turn becomes the source of unequal distribution of the benefits from the new developments as well. The two examples discussed in Chapter V of successful reforestation on a country-wide basis—South Korea and China—while vastly different in their political contexts, have at least one significant feature in common: in both cases a radical agrarian reform programme had already been undertaken which had established relative equality in the country's agrarian structure. In this context, it is noteworthy that even within a noted overall egalitarian land ownership pattern in South Korea in the 1970s (and in spite of the feasibility of top-down State pressure), in villages where the land distribution continued to be relatively unequal, attempts to promote community participation met with difficulties. All said, where inequalities are sharp, redistributive land reform may well prove to be a necessary pre-condition for community reforestation schemes to take root.

Again when we consider the issue of the organisational structure of R and D and extension networks, or the attitudes of the individuals who comprise these networks, a more fundamental change is likely to be needed than a mere 're-training' of these individuals. One aspect again concerns people's relative material circumstances and the

*'Make the earth a garden'*

difficulty of bringing about a two-way interaction between the economically well-off and the poor. But additionally there is the question of entrenched 'social status' considerations and the psychological distance between those involved in physical versus mental labour, between the town dwellers and the villagers, or between the sexes, which relates to more than just differences in material circumstances. It makes itself felt both in the way schemes are implemented (i.e. the extent of user-involvement) and in the way the programmes are evaluated (i.e. how the evaluator relates to the villagers).

Following these implications to their logical conclusion, it can be seen that attempts to bring about some fundamental restructuring towards equality in the social and economic base of many Third World countries, as well as changes at the ideological level, would need to be a necessary part of efforts to alleviate the woodfuel crisis, if such efforts are to make any significant impact.

At the same time, it is also apparent that such a restructuring is typically unlikely to be initiated from above—from the sources of power and privilege; and that the pressure and challenge to change will need to come from below—from grassroots organizations of the underprivileged. It is in this context that the ongoing struggles for economic and social equality (including tribal struggles against their alienation from forest land) in different parts of the Third World, or people's movements for saving trees and for reforestation, or even the existing localized efforts of some voluntary groups to promote rural development schemes through a more participatory and 'people oriented' approach, constitute important thrusts; although how they will or can come together, to provide a wider-based challenge for change, continues to be an issue of ongoing controversy and debate.

# APPENDIX

# THE DIFFUSION OF IMPROVED CHARCOAL KILNS

Wood can be converted into a whole range of secondary fuels such as charcoal, producer gas, water gas, methyl alcohol, etc. The conversion from wood to charcoal can be undertaken through kilns and furnaces, or through pyrolytic equipment such as retorts; the latter also enable the capture of woodfuels other than charcoal. However, as noted in Chapter II, it is the use of woodfuel in the form of charcoal which is of primary interest here. This is because after firewood itself charcoal is the most commonly used woodfuel in Third World countries; and, further, secondary woodfuels other than charcoal essentially serve industrial purposes. For example, producer gas can energize internal combustion engines (Earl, 1975: 36), and the problems relating to the diffusion of hardware that produces these fuels are not likely to be especially different from those relating to other small-scale industrial equipment. Hence this section is essentially concerned with the diffusion of improved kilns for producing charcoal.

In this context certain aspects of charcoal use and production need highlighting. Existing literature, though limited, indicates that unlike firewood which is largely collected by rural households for their own use, charcoal is produced primarily for sale to the urban areas, and is bought largely by the urban rich. Some instances are observed in Africa (e.g. in Ghana and Malawi) where families produce small amounts of charcoal for their own use, but this is noted to account for only a small part of total charcoal production in the region (see DEVRES, 1980: 70).

Literature on who the charcoal producers are, is again sparce. Most studies dealing with charcoal production have devoted space essentially to the technical characteristics of the hardware in use, and to the relative efficiencies of different conversion processes;[1] few have described the socio-economic characteristics of the people making charcoal.

These few provide significant pointers, although the situation could be somewhat different between countries. *Sylva Africana* (1980) a quarterly newsheet issued by the Forestry Co-operative Research Project at Nairobi (Kenya) notes, for example, that in some parts of Africa charcoal is produced in earth kilns as a kind of cottage industry run by small farmers, either individually or in small groups. When ready, it is put in sacks and carried to the nearest road for sale to passing vehicles. In other areas, for many people, charcoal production represents a full-time occupation or at least a dry-season one; here lorries often enter the forest and pick up loads from production areas. In most African countries licences are needed for production and transportation but typically producers tend to be unlicensed.

This observation is supported by Kabagambe's (1976) study of charcoal production in Kenya. He notes that charcoal for domestic purposes (the bulk of which is for urban consumption) tends to be produced, on the one hand, by a few big producers and, on the other, by numerous small ones. The former operate in controlled forest reserves, are licensed, and have to pay 'royalties' to the government. They sell directly to industry or to charcoal dealers—wholesalers— who in turn sell to retailers in towns. The small producers are generally unlicensed and operate in uncontrolled forest areas. They produce small quantities of output which they usually take to the roadside, from where it is collected by the dealers and sold in the towns. Some of the output of the small producers also goes to supply rural domestic needs or rural trading centres. The channels of distribution here are informal and the product does not pass through institutionalized marketing systems. Kabagambe does not give any estimate or even 'guesstimate' of the proportion of total charcoal that comes from the big producers relative to the small, nor how many producers there are likely to be. But this is not surprising since the unregulated nature of the operations would make it extremely difficult to obtain such information.

A study by Hoskins (1979b), discussed in more detail further on, likewise supports the view that there tend to be a large number of small producers, and she provides the interesting detail that they are sometimes hired by the dealers who take charge of the transportation and sale functions. Brokensha and Riley (1978) again note in the context of the Mbere district of Kenya that most of the charcoal makers are poor people for whom this is the main source of livelihood. DEVRES (1980: 70) similarly observes that in many parts of Latin America such as in Mexico, Brazil and Guatemala charcoal makers are almost exclusively the poor. Usually charcoal making is taken to be a 'poverty sign' and in some cultures is viewed as 'dirty' work to be shunned by the well-off.

Not surprisingly, therefore, earth kilns which require no financial investment continue to be the most commonly used means of converting wood to charcoal. This was noted in most studies including Powell's (1978: 119-120) study of charcoal production in Ghana, where he estimates that 99 per cent of the charcoal is produced in this way. The conversion efficiency of earth kilns, however, is estimated to be quite low—varying between 5 to 15 per cent, depending on the dryness of the wood, relative to an estimated conversion efficiency of 25 per cent or more of improved kilns (Powell, 1978: 123). Hence the use of improved technology is undoubtedly desirable.

Of course, improved charcoal kilns, unlike improved wood-burning stoves, cannot be seen as a direct means of alleviating the woodfuel crisis of the rural population. They could, however, have indirect implications in that more efficient production would augment the income of those among the poor who produce charcoal, and it could help reduce pressure on forest resources insofar as improved kilns can make use of small timber which would otherwise lie unused in the forest, and which cannot be used in the earth kilns.

What problems might be expected in the diffusion of such kilns? Unfortunately there appear to be hardly any case studies which have looked at the actual experience of kiln diffusion (the study by Hoskins, discussed further on, being one of the few exceptions). Hence, we have to depend a good deal on *a priori* reasoning. On *a priori* grounds we can say that unlike wood-stoves, in the case of charcoal kilns the problem of need-perception is less likely to arise because charcoal is largely sold—hence a shift to efficient production will bring obvious financial benefits to the adopter, and not just benefits in terms of saving women's and children's labour time, smokelessness, etc. Further, kilns relate more to the male domain of activity than the female; hence the problem of the men not perceiving the advantages of an innovation because it benefits mainly the women, as with wood-stoves, is less likely to be applicable.

Exceptions to this do exist. For example, Fleuret and Fleuret (1978) note in their sample survey relating to Tanzania, that women, especially in the poor households, spend a substantial part of their time producing charcoal for sale. However, the bulk of the literature does suggest that charcoal production generally tends to be a male occupation or a family based one, rather than an exclusively female one.

The main difficulties with diffusion, however, are likely to be of a different nature. What the studies point to is that charcoal-making is still not a large-scale commercial activity. Insofar as the producers are small in scale but large in number, the introduction of improved technology requires reaching a mass of producers, except in those cases where the producers have been able to organize themselves on a cooperative basis, or could be organized on a cooperative basis. Let us consider the likely problems with each of these two possibilities.

Perhaps the trickiest problem of reaching a mass of small charcoal producers lies in the unregulated, and often illegal nature of their activity; many are not likely to admit undertaking it, at least not admit it to official extension agents. Added to this is the fact that the small producers are often very poor, so that even assuming that they can be located and identified through informal channels, they would find even small monetary investments difficult: earth kilns, like open fires, require no financial outlay, whereas a shift to improved kilns is likely to require some cash expenditure. If the immediate personal resources of the producer are insufficient to provide the cash, the availability of institutional credit again becomes a crucial factor; and all the shortcomings and biases in the workings of rural credit institutions mentioned earlier would be applicable here as well. The possibilities of the diffusion of improved kilns under these circumstances need to be considered in the context of producers' cooperatives or of adoption by the 'big' producers. According to the *Sylva Africana* (1980) some cottage industry type concerns run by groups of farmers do exist in parts of Africa. These offer scope both for extending credit to buy improved kilns and for cooperation with kiln designers, so that designs can be generated and adapted to user requirements and specifications. Big producers could likewise be reached through the usual productive-technology-diffusing channels.

But to reach the large number of small producers still operating independently, it would be necessary to go beyond this if some sort of dent has to be made, that is, it would be necessary to promote producers' cooperatives. Insofar as the producers are 'equal' in their smallness, their individual interests would not be antagonistic to one another, and at face value it could be expected that many of the problems that tend to arise if cooperation is desired between those belonging to different levels of the socio-economic hierarchy, would not apply here.

In practice, however, the problem is likely to be more complex, since the charcoal producers are linked with the wider community through a set of intermediaries—in this case the dealers—who make a profit out of the unorganized nature of production. Operating within a cooperative, presumably licensed, and possibly with their own transportation and marketing arrangements, the earlier small producers would no longer be so 'small' or vulnerable or so open to exploitation. In these circumstances it would be unrealistic to expect the middlemen, who would lose out in this deal, to remain passive, and the success of the venture could depend crucially on the degree of power that these middlemen can continue to exercise.

Some of these features are in fact highlighted in the description by Hoskins (1979b: 28-30) of an attempt to diffuse improved charcoal-producing technology in an African country (she does not name it). In fact hers is perhaps the only case study (or at least one of very few) which considers the diffusion experience in some detail. Powell (1978), who also speaks of an attempt to introduce an improved kiln (the Tranchant kiln) in Ghana, while noting that the attempt failed because the kiln was too expensive and needed skilled operators, does not provide any additional information on who the producers were and in what social context they were operating.

The programme Hoskins describes was designed as a two-stage experiment. In the first stage a kiln was to be developed which would make more complete use of trees, including odd-shaped pieces of wood, and convert them to charcoal more efficiently than the earth kiln. In the second stage the aim was to form cooperatives of charcoal makers, eliminate middlemen and ensure that the cooperatives used wood-conserving techniques.

The programme was launched by the project designer with the help of the forestry service, and on its initiation a variety of restrictions were imposed on the activities of the traditional charcoal makers and middlemen (who hire charcoal makers and control the transport and sale aspects). The restrictions included the issue of licences to a limited number of middlemen; placing a limit on the number of charcoal makers who could work under each middleman; and restricting the areas in which wood could be collected.

While these measures helped to reduce wood consumption, they also left several hundred people unemployed. Further, though the design experiments for improving the kiln were done with the help of some of the charcoal producers and middlemen, the charcoal so produced was sold at subsidized rates, thus underselling the traditional vendors. The many middlemen and charcoal producers who thus became unemployed because of the licensing policy were highly resentful, and several were reported to have started forest fires in the protected areas. Those holding licences before the project was introduced were also resentful since they now had to face competition from the subsidized charcoal produced on the project. Hence though the design part of the experiment appears to have been successful—the technicians had designed a relatively inexpensive kiln which accommodated various sized wood pieces—the same cannot be said of the project as an experiment in diffusion.

There are a number of reasons why the project as conceived was unlikely to work in these terms:

(i) While the new kiln was cheap and economized on wood it still had to compete with 'a mould of clay' (the earth kiln) that needed no expenditure. Hoskins does not mention the provision of credit as being part of the scheme.

(ii) The demand for new kilns did not come from the charcoal-makers themselves who had not experienced a shortage of wood until the new restrictions were imposed; hence ensuring cooperation was problematic.

(iii) The new licensing policy was discriminatory since it did not cover all the small producers; nor did it provide any alternative sources of livelihood to those left out.

(iv) Ultimately, the reason for seeking to eliminate the middlemen was to prevent them from exploiting the small producers. However the way in which this was attempted suggests an inadequate understanding by those instituting the project of the social organisation of charcoal production in the area. This is reflected in their failure to take into account two related aspects: first the likelihood of the charcoal producers being tied to the middlemen through a variety of economic and extra-economic coercive relationships that cannot be dissolved merely by decree; and second the expected resentment and reactions of the middlemen and producers who were left out of the scheme, and who were therefore likely to attempt thwarting the project in various ways—as did in fact happen.

Basically, despite the involvement of some of the local people in kiln design, in its overall implementation the project reflects the familiar top-down approach, in addition to its being simplistic in its conception of the political economy aspects of technical change and diffusion processes.

# REFERENCES

Abdullah, T.A., and S.A. Zeidenstein (1978): 'Finding Ways to Learn about Rural Women: Experiences from a Pilot Project in Bangladesh', *Sociologia Ruralis*, Vol. XVIII, No. 2/3.

Acharjee, B., Chatterjee, R., and Chanda, S.K. (1974a): 'Efficiency of Domestic *Chullah*', *Indian Standards Institution Bulletin*, Vol. 26, January.

Acharjee, B., S.K. Chanda, and R. Chatterjee, (1974b): 'Efficiency of Masonry Domestic Chullah', *Indian Standards Institution Bulletin*, Vol. 26, September.

Acharya, Meena and Lynn Bennett (1981): *The Rural Women of Nepal: An Aggregate Analysis and Summary of Eight Village Studies*, in *The Status of Women in Nepal*, Vol. II, Part 9, CEDA, Tribhuvan University, Nepal.

Adeyoju, S.K., (1976): 'Land Tenure Problems and Tropical Forestry Development', mimeo, Committee on Forest Development in the Tropics, Fourth Session, Rome, Italy, 15-20 November.

Agarwal, Anil (1983a): 'The Cooking Energy Systems—Problems and Opportunities', discussion paper, Center for Science and Environment, Delhi.

Agarwal, Anil (1983b): 'In the Forests of Forgetfulness' *The Illustrated Weekly of India*, November 13-19, Bombay.

Agarwal, Anil and Priya Deshingkar (1983): 'Headloaders: Hunger for Firewood—I', CSE Report No. 118, Center for Science and Environment, Delhi.

Agarwal, Anil and Bhubanesh Bhatt (1983): 'Firewood in Cities—I: The Dimension of the Rural-Urban Firewood Trade', CSE Report No. 112, Center for Science and Environment, Delhi.

Agarwal, Bina (1981): 'Agricultural Modernisation and Third World Women: Pointers from the Literature and on Empirical Analysis', World Employment Programme Research Working Paper No. WEP 10/WP 21, International Labour Office, Geneva.

Agarwal, Bina (1984): 'Rural Women and the High Yielding Variety Rice Technology', *Economic and Political Weekly*, Review of Agriculture, Vol. 19, No. 13. March.

Agarwal, Bina (1985): 'Women and Technological Change in Agriculture: Asian and African Experience' in *Technology and Rural Women: Conceptual and Empirical Issues*, ed. by Iftikar Ahmed, George Allen and Unwin, London.

Agarwala, V.P., D.W. Seckler and K.G. Tejwani (1982): *Energy from Trees: The Second Green Revolution in India*, Society for Promotion of Wastelands Development, Delhi.

Aguilar, Filomeno V. Jr. (1982): *Social Forestry for Upland Development: Lessons from Four Case Studies*, Institute of Philippine Culture, Ateneo de Manila University, Quezon City, Philippines.

Alvares, Claude (1984): 'From Chipko to Appiko', *Indian Express*, Sunday Magazine, India, September 9.

Apthrope, R.J. (ed., 1970): *Rural Cooperatives and Planned Change in Africa: Case*

*Materials*, United Nations Research Institute for Social Development, Geneva.

Arens, Jenneke, and Jos Van Beurden (1977): *Jhagrapur: Poor Peasants and Women in a Village in Bangladesh*, Amsterdam.

Arnold, J.E.M., and J.H. Jongma (1977): 'Fuelwood and Charcoal in Developing Countries', *Unasylva*, FAO, Vol. 29, No. 118.

Arnold, J.E.M. (1978): 'Wood Energy and Rural Communities', mimeo, Forestry Department, FAO, Rome: paper presented at the 8th World Forestry Congress, Jakarta, Indonesia, October.

Arnold, John H. Jr. (1980): 'A Revised Methodology for Energy Demand Surveys', discussion paper written for the Workshop on Energy Assessment Methodologies. The Board on Science and Technology for International Development, National Academy of Sciences, Washington D.C.

Arrow, K.J (1963): *Social Choice and Individual Values*, 2nd edition, Wiley, New York.

ASTRA (1981): 'Rural Energy Consumption Patterns: A Field Study', Indian Institute of Science, Bangalore.

Avila, Charlie (1976): *The Peasant Theology*, World Student Christian Federation, Asia; Bangkok, Thailand.

Bahuguna, Sunderlal (1984): 'Women's Non-violent Power in the Chipko Movement', in *In Search of Answers: Indian Women's Voices from Manushi*, ed. by Madhu Kishwar and Ruth Vanita, Zed Books, UK.

Bahuguna, Sunderlal (n.d.): 'My Experience of Eucalyptus', mimeo, Parvatiya Navjivan Mandal, Tehri-Garhwal, UP, India.

Bajracharya, Deepak (1979): 'Firewood Consumption in the Nepal Hills: A Comparison of the Annual Memory Recall and the Weekly Recorded Data', paper presented at the Conference on Rapid Rural Appraisal, Institute of Development Studies, University of Sussex, UK, 4-7 December.

Bajracharya, Deepak (1981): *Implications of Fuel and Fodder Needs for Deforestation: An Energy Study in a Hill Village Panchayat of Eastern Nepal*, D. Phil. dissertation, Science Policy Research Unit, University of Sussex, UK.

Bajracharya, Deepak (1983a): 'Fuel, Food and Forest: Dilemmas in a Nepali Village', Working Paper WP-83-1, Resource Systems Institute, East-West Center, Honolulu, Hawaii.

Bajracharya, Deepak (1983b): 'Deforestation and the Food/Fuel Context: Historico-Political Perspectives from Nepal', Working Paper, Resource Systems Institute, East-West Center, Honolulu, Hawaii.

Ball, Mike (1979): 'Cost-Benefit: A Critique', in *Issues in Political Economy: A Critical Approach*, ed. by Francis Green and Peter Nore, the Macmillan Press Ltd., London.

Bandhyopadhyay, Jayanto (1981): 'Beyond the Firewood March', *Financial Express*, 1st and 2nd September, India.

Bandhyopadhyay, Jayanto and Vandana Shiva (1984): 'Growing Wisely or Growing Well?' *Indian Express*, Sunday Magazine, July 15, India.

Bapat, Shailaja (1983): 'It Lacks Vitality', *The Economic Times*, India, August 6.

Barker, D., J. Oguntoginboj and P. Richards (1977): 'The Utility of the Nigerian Peasant Farmers' Knowledge', in the Management, Monitoring and Assessment Research Center, Chelsea College, London University.

Barker, D. and P. Richards (1978): 'Repertory Grid Methods and Environmental Images in Rural Africa', Institute of British Geography, School of Oriental and

African Studies, March.

Barker, David (1979): 'Appropriate Methodology: An Example using a Traditional African Board Game to Measure Farmer's Attitudes and Environmental Images', in *Rural Development: Whose Knowledge Counts? IDS Bulletin*, Vol. 10, No. 2, January, Institute of Development Studies, University of Sussex, UK.

Barnett, Andrew, Leo Pyle and S.K. Subramaniam (1978): *Biogas Technology in the Third World: A Multidisciplinary Review*, Report 103c, International Development Research Center, Ottawa, Canada.

Barnett, Andrew, Martin Bell and Kurt Hoffman (1982): *Rural Energy and the Third World*, Pergamon Press, UK.

Batliwala, Srilata (1983): 'Woman and Cooking Energy', *Economic and Political Weekly*, Vol. 18, Nos. 52 and 53, December 24-31.

Bell, Martin (1979): 'The Exploitation of Indigenous Knowledge or the Indigenous Exploitation of Knowledge: Whose Use of What for What?', in *Rural Development: Whose Knowledge Counts? IDS Bulletin*, Vol. 10, No. 2, January, Institute of Development Studies, UK.

Bernales, Benjamin C. and Angelito P. Dela Vega (1982): *Case Study of Forest Occupancy Management Programme in Dona Remedios, Trinidad, Bulacan*, Integrated Research Center, Dela Salle University, Taft Avenue, Manila, Philippines.

Bertocci, Peter J. (1975): 'The Position of Women in Rural Bangladesh', Paper No. 33, International Seminar on Socio-Economic Implications of Introducing HYVs in Bangladesh, Bangladesh Academy for Rural Development, Comilla, Bangladesh, April 9-11.

Bhaduri T. and V. Surin (1980): 'Community Forestry and Women Headloaders' in *Community Forestry and People's Participation—Seminar Report*, Ranchi Consortium for Community Forestry, November 20-22.

Bharara, L.P. and M.L.A. Sen (1970): 'Social Aspects of Farm Forestry in the Arid Zone', *Annals of Arid Zones*, Vol. 9, No. 1, March, Arid Zone Research Association of India, Jodhpur.

Bhasin, Kamla (1976): *Participatory Training for Development*, Report of the FFHC/AFD, FAO, Bangkok, Thailand.

Bhasin, Kamla (1984): 'Are We on the Right Track? Some Thoughts on Participatory Self-evaluation', mimeo, FFHC/AFD, Food and Agricultural Organisation, India.

Bhatia, Ramesh (1980): 'Energy and Rural Development: An Analytical Framework for Socio-Economic Assessment of Technological and Policy Alternatives', working paper, Resource Systems Institute, East-West Centre, Honolulu, Hawaii, January.

Bialy, Jan. (1979a): 'Energy Flow in a Rural Society: A Case Study from Sri Lanka', precis of thesis, School of Engineering Science, University of Edinburgh, England.

Bialy, Jan. (1979b): 'Measurement of the Energy Released in the Combustion of Fuels' occasional papers on Appropriate Technology, ATO22, School of Engineering Science, University of Edinburgh, England, December.

Biggs, Stephan D. (1980): 'Research by Farmers: The Importance of the Informal Agricultural R and D System in Developing Countries', mimeo, Institute of Development Studies, University of Sussex, UK, February.

Biggs, Stephan D. and Edward J. Clay (1981): 'Sources of Innovation in Agricultural Technology', *World Development*, Vol. 9, No. 4, April.

Blanchet, Therese (1982): 'Becoming an Insider'—Women's Privileges in Village Fieldwork', in *Exploring the Other Half—Field Research with Rural Women in*

*Bangladesh*, ed. by Shamima Islam, Women for Women, Dhaka, Bangladesh.

Bokil, K.K. and K.S. Rao (1979): 'Energy Plantations in the Semi-Arid Rural Regions of Saurashtra and Kutch', paper presented at the Research Planning Workshop on Energy and Rural Development, Institute of Management, Ahmedabad, India, December 27-29.

Bose, A.B. and L.P. Bharara (1965): 'Some Sociological Considerations in Farm Forestry', *Annals of Arid Zones*, Vol. 4, No. 1, March.

Boserup, Ester (1970): *Women's Role in Economic Development*, St. Maitin's Press, New York.

BOUIES (1972): *Bulletin of the Oxford University Institute of Economics and Statistics*, Vol. 34, February.

BRAC (1980): *The Net: Power Structure in Ten Villages*, Bangladesh Rural Advancement Committee, Dhaka, Bangladesh.

Brandt, Vincent S.R. and Man-gap Lee (1980): 'Community Development in South Korea (1971-1976)', mimeo, forthcoming in *Community Development in the 1970s* ed. by R.P. Dore, IDS, University of Sussex, UK.

Briscoe, John (1979): 'Energy Use and Social Structure in a Bangladesh Village', *Population and Development Review*, Vol. 5, No. 4, December.

Brokensha, David and Bernad Riley (1978): *Forestry, Foraging, Fences and Fuel in a Marginal Area of Kenya*, Social Process Research Institute, Santa Barbara, University of California.

Brokensha, David and Bernard W. Riley (1980): 'Mbeere Knowledge of their Vegetables and its Relevance for Development: A Case Study from Kenya', in *Indigenous Knowledge Systems and Development*, ed. by David Brokensha, D.M. Warren and Oswald Werner, The University Press of America, Lanham.

Brown, Norman L. (ed., 1978): *Renewable Energy Resources and Rural Applications in the Developing World*, AAAS Selected Symposia Series No. 6, Westside Press Inc., USA.

Bukh, Jette (1979): *Village Women in Ghana*, Center for Development Research, Scandinavian Institute of African Studies, Uppsala.

Burley, Jeffery (1978): 'Selection of Species for Fuelwood Plantations', paper presented for the 8th World Forestry Congress, Jakarta, Indonesia, 16-28 October.

Burton, H. (1969): *Exmoor*, Hodder and Stoughton, London.

Byres, T.J. (1972), 'The Dialectics of India's Green Revolution', *South Asian Review*, London, January.

Cain, Mead, Syeda Rokeya Khanam and Shamsun Nahar (1979): 'Class, Patriarchy, and the Structure of Women's Work in Rural Bangladesh', Working Paper No. 43, Center of Population Studies, The Population Council, New York, May.

Cecelski, Elizabeth (1984): 'The Rural Energy Crisis, Women's Work and Family Welfare: Perspectives and Approaches to Action', World Employment Programme Research Working Paper No. WEP 10/WP 35, International Labour Office, Geneva.

Cernea, Michael M. (1981): 'Land Tenure Systems and Social Implications of Forestry Development Programmes', World Bank Staff Working Paper No. 452, April.

Chakravarty, Kumaresh and G.C. Tiwari (c. 1977): 'Regional Variation in Women's Employment: A Case Study of Five Villages in Three Indian States', mimeo, Programme of Women's Studies, Indian Council for Social Science Research (ICSSR), New Delhi.

Chambers, Robert (1983): *Rural Development—Putting the Last First*, Longman Group Ltd., London.

Chand, Malini and Rekha Bezboruah (1980): 'Employment Opportunities for Women in Forestry', in *Community Forestry and People's Participation—Seminar Report*, Ranchi Consortium for Community Forestry, November 20-22.

*Chipko-Geet* (1979): 'Songs of the Chipko Movement', Parvatiya Navjivan Mandal, Ghansali, Tehri-Garhwal, UP, India.

Chlala, Henry G. (1972): 'The Present Situation and Future Prospects of Charcoal Production and Consumption in Kenya', mimeo, Industrial Survey and Promotion Center, Ministry of Commerce and Industry, Nairobi.

Chowdhury, Kamla (1982b): 'The Greening of India', *Indian Express*, India, 21 December.

Chowdhury, Kamla (1983): 'Schools as Partners in Social Forestry', Discussion Paper Series, DP No. 11, Ford Foundation, Delhi, August.

Chowdhury, Neerja (1982a): 'Forest Policy—New Bill for Whom?' *Voluntary Action*, May 1982.

Chung, Jae Won (1979): *Biogas in Korea*, Final Report of Social and Economic Evaluation on Biogas Technology in Korea, Office of Rural Development, Suweon, Korea, May.

Clay, Edward J. (1980): 'The Economics of the Bamboo Tubewell: Dispelling Some Myths about Appropriate Technology', *CERES* (FAO), Vol. 13, No. 3, May-June 1980.

Cohen, John M. and Norman Uphoff (1980): 'Participation's Place in Rural Development: Seeking Clarity through Specificity', *World Development*, Vol. 8, March.

Consortium for International Development (1978): *Proceedings and Papers of the International Conference on Women and Food, Vol. III*, University of Arizona, Tucson, Arizona, January 8-11.

Cooper, Charles (1979a): 'A Summing Up of the Conference', in *Appropriate Technologies for Third World Development*, ed. by Austin Robinson, Proceedings of a Conference held by the International Economic Association at Teheran, Iran, Macmillan Press Ltd., London.

Cooper, Charles (1979b): *Economic Evaluation and Environment—A Methodological Discussion with Particular Reference to Developing Countries*, prepared for United Nations Environment Programme, Institute of Development Studies and Science Policy Research Unit, University of Sussex, UK, forthcoming as a book, to be published by Hodder and Stoughton, London.

CWDS (1984): 'Role and Participation of Women in the *Chipko* Movement in the Uttarkhand Region in Uttar Pradesh, India', World Employment Programme Working Paper No. WEP 10-4-04-18-1, draft, ILO, Geneva.

Dasgupta, Ajit K. and D.W. Pearce (1972): *Cost-Benefit Analysis: Theory and Practice*, Macmillan Press Ltd., UK.

Dasgupta, Biplab (1977): *Agrarian Change and the New Technology in India*, Report No. 77.2, United Nations Research Institute for Social Development, Geneva.

Dean, Genevieve C. (1972): 'Science, Technology and Development: China as a Case Study', *The China Quarterly*, London, July-September.

DEVRES (1980): *The Socio-Economic Context of Fuelwood Use in Small Rural Communities*, AID Evaluation Special Study No. 1, USAID, August.

Desai, Ashok V. (1980): *India's Energy Economy: Facts and their Interpretation*,

Economic Intelligence Services, Centre for Monitoring Indian Economy, Bombay, India, February.

Digerness, Turi Hammer (1977): 'Wood for Fuel—The Energy Situation in Bara, the Sudan', mimeo, Department of Geography, University of Bergen, Norway, July.

Digerness, Turi Hammer (1979): 'Fuelwood Crisis Causing Unfortunate Land Use and the Other Way Around', *Norsk. Geogr. Tidsskr.*, Oslo, Vol. 33.

Dixon, Ruth B. (1978): *Rural Women and Work: Strategies for Development in South Asia*, John Hopkins University Press, Baltimore and London.

D'Monté, Darryl (1982): 'Pine for the forest?' *The Hindustan Times*, January 4.

Dogra, Bharat (1984), *Forests and People*, Bharat Dogra, A-2/184, Janakpuri, New Delhi.

Dommen, Arthur J. (1975): 'The Bamboo Tubewell: A Note on an Example of Indigenous Technology', *Economic Development and Cultural Change*, Vol. 23, No. 3, April.

Douglas J. Sholto and Robert A. de J. Hart (1976): *Forest Farming: Towards a Solution to Problems of World Hunger and Conservation*, Robinson and Watkins Press, London.

Draper, S.A. (1977): 'Wood Processing and Utilisation at the Village Level', Third FAO/SIDA Expert Consultation on Forestry for Local Community Development, Semarang, Indonesia, December.

Draper, S.A. (1977): 'Wood Processing and Utilisation at the Village Level', Third FAO/SIDA Expert Consultation on Forestry for Local Community Development, Semarang, Indonesia, December.

Dumont, René (1973): 'A Self-Reliant Rural Development Policy for the Poor Peasantry of Sonar Bangladesh', Ford Foundation, Dhaka, May.

Dutt, G.S. (1978a): 'Reducing Cooking Energy Use in Rural India', Report PU/CES 74, Center for Environmental Studies, Princeton University, Princeton, November 21.

Dutt, G.S. (c. 1978b): 'Efficient Wood Burning Cooking Stove Literature', draft mimeo, Center for Environmental Studies, Princeton University, Princeton.

Dutt, G.S. (c. 1978c): 'Improved Wood Burning Cooking Stoves for LDCs', mimeo, Center for Environmental Studies, Princeton University, Princeton.

Earl, D.E. (1975): *Forest Energy and Economic Development*, Clarendon Press, Oxford.

Earthscan (1983): 'Earthscan Press Briefing Document No. 37', mimeo, International Institute for Environment and Development, London.

Eckholm, Erik.P. (1975): 'The Other Energy Crisis', World Watch Paper No. 1, World Watch Institute, USA, September.

Eckholm, Erik P. (1976): *Losing Ground: Environmental Stress and World Food Prospects*, World Watch Institute, USA.

Eckholm, Erik P. (1979): 'Planting for the Future: Forestry for Human Needs', World Watch Paper No. 26, World Watch Institute, USA, February.

Eckholm, Erik P. (1980): 'Introduction in Firewood Crops—Shrubs and Tree Species for Energy Plantations', National Academy of Sciences, Washington D.C.

Ernst, Elizabeth (1977): 'Fuel Consumption Among Rural Families in Upper Volta, West Africa', mimeo, Ouayadougou, Peace Corps, July 5.

Evans, I (1978): 'Using Firewood More Efficiently', paper prepared for the 8th World Forestry Congress, Jakarta, Indonesia, 16-28 October.

Evans, David and James Hoxeng (c. 1973): 'The Ecuador Project', Technical Note No. 1, the Ecuador Non-formal Education Project, Ministry of Education, Ecuador, and the Center for International Education at the University of Massachusetts, USA.

FAO (1977): Report on the FAO/SIDA Expert Consultation in Forestry for Community Development, held at Rome, Italy, 21-22 June.

FAO (1978a): 'Energy and Agriculture' in *The State of Food and Agriculture 1976*, Food and Agricultural Organisation, Rome.

FAO (1978b): Report on the Third FAO/SIDA Expert Consultation on Forestry for Local Community Development, held at Semarang, Indonesia, 5-15 December.

FAO (c. 1979): *Forestry for Rural Communities*, Forestry Department, Food and Agricultural Organisation, Rome, Italy.

FAO (1981): *Map of the Fuelwood Situation in Developing Countries—Explanatory Note*, Food and Agricultural Organisation, Rome.

FAO (1982a): *Village Forestry Development in the Republic of Korea*, Food and Agricultural Organisation, Rome.

FAO (1982b): *Forestry in China*, Food and Agricultural Organisation, Rome.

FAO (1984): *1982 Yearbook on Forestry Products*, FAO Forestry Statistics Series, Food and Agricultural Organisation, Rome.

Feuerstein, Marie Theresé (1979): 'Establishing Rapport' paper presented at a conference on Rapid Rural Appraisal, Institute of Development Studies, University of Sussex, UK, 4-7 December.

Fleuret, Patrick, C. and Anne Fleuret (1978): 'Fuelwood Use in a Peasant Community: A Tanzanian Case Study', *The Journal of Developing Areas*, 12, April.

Floor, W.M. (1977): 'The Energy Sector of the Sahelian Countries' mimeo, Policy Planning Section, Ministry of Foreign Affairs, The Netherlands, April.

Foley, Gerald and Patricia Moss (1983): *Improved Cooking Stoves in Developing Countries*, Earthscan, Technical Report No. 2, IIED, London.

Folger, Bonnie and Meera Dewan (1983): 'Kumaon Hills Reclamation: End of Year Site Visit', mimeo, OXFAM America (Delhi office).

Fortmann, Louise and Dianne Rochcleau (n.d.): 'Women and Agro-Forestry: Four Myths and Three Case Studies', draft paper, Department of Forestry and Natural Resources, University of California, Berkeley.

French, David (1978): 'Renewable Energy for Africa: Needs, Opportunities, Issues', paper submitted to USAID, Washington D.C., AFR/DR/SDP, July 14.

French, David (1979): *The Economics of Renewable Energy Systems for Developing Countries*, USAID, Washington D.C., January.

Gaikwad, V.R., B.L. Tripati and G.S. Bhatnagar (1972): *Opinion Leaders and Communication in Indian Villages*, Centre for Management of Agriculture, Indian Institute of Management, Ahmedabad, India.

Gamser, Mathew S. (1979): 'The Forest Resource and Energy Development in the Third World', M.Sc. dissertation, HSSS, University of Sussex, UK, July.

Goldemberg, José (1978): *Energy Strategies for Developed and Less Developed Countries*, Report PU/CES 70, The Center for Environmental Studies, Princeton University, February 10.

Government of India (1948): 'India's Forests and the War', Ministry of Agriculture, Delhi.

Government of India (1964): *A Study on the Efficiency of Chulahs*, National Buildings

Organisation and United Nations Economic Commission for Asia and the Far East, Regional Housing Center.

Government of India (1980): *Proceedings of the Seminar on the Role of Women in Community Forestry*, Ministry of Agriculture, Department of Agriculture and Cooperation, India.

Government of India (1982): *Report of the Fuelwood Study Committee*, Planning Commission, Government of India, March.

Griffin, Keith (1971): *The Green Revolution: An Economic Analysis*, United Nations Research Institute for Social Development, Report No. 72.6, Geneva.

Griliches, Zvi (1957): 'Hybrid Corn: An Exploration in the Economics of Technological Change', *Econometrica*, Vol. 25, October.

Griliches, Zvi (1960): 'Hybrid Corn and the Economics of Innovation', *Science*, 132.

Guha, Ramachandra (1983): 'Forestry in British and Post-British India—A Historical Analysis', *Economic and Political Weekly*, October 29.

Gujarat Forest Department (1984): *Eucalyptus Farming: Facts and Fears*, The Government Press, Vadodara, India.

Guleria, Amar and Tirath Gupta (1982): *Non-wood Forest Products in India : Economic Potentials*, Oxford and IBH, India.

Gupta, Krishnamurthi and Deshbandhu (eds., 1979): *Man and Forest: A New Dimension in the Himalayas*, Today and Tomorrow Printers and Publishers, India.

Hanger, Jane and Jon Moris (1973): 'Women and the Household Economy', in *Mwea: An Irrigated Rice Settlement in Kenya*, ed. by Robert Chambers and Jon Moris, Afrika Studien No. 83, Ifo-Institut fur Wirschaftsforchung Muchen, Afrika Studien Stelle, Muchen.

Hapgood, David (1965): *Policies for Promoting Agricultural Development*, Report of a Conference on Productivity and Innovation in Agriculture in the Underdeveloped Countries, Report No. C/65-3, Center for International Studies, MIT, Cambridge , USA.

Hapgood, David (1968): 'The Politics of Agriculture', *Africa Report*, November.

Harjosoediro, Soedarwano (1979): 'Forestry Under Conditions of Population Pressure in Indonesia', in *Forestry in National Development: Production Systems, Conservation, Foreign Trade and Aid*, ed. by K.R. Shephard and H.V. Richter, Development Studies Center, Australian National University, Monograph No. 1.

Harrison, S. Brown and Kirk R. Smith (1980): 'Energy for the People of Asia and the Pacific', draft, section on Rural Energy, Resource Systems Institute, East-West Center, Honolulu, Hawaii; also in *Annual Review of Energy*, Vol. 5, 1980.

Hayami, Yujiro and V.W. Ruttan (1971): *Agricultural Development: An International Perspective*, The John Hopkins Press, Baltimore, Maryland.

Hayes, Dennis (1977): *Energy for Development: Third World Options*, World Watch Paper No. 15, World Watch Institute, Washington D.C.

Herrera, A. (1975): 'Scientific and Traditional Technologies in Developing Countries', mimeo, Science Policy Research Unit, University of Sussex, UK.

Herrera, A (1981): 'The Generation of Technologies in Rural Areas', *World Development*, Vol. 9, No. 1, January.

Hoffman, H.K. (1980): 'Alternative Energy Technologies and Third World Rural Energy Needs: A Case of Emerging Technological Dependency', *Development and Change*, July.

Hoskins, Marilyn (1979a): 'Women in Forestry for Local Community Development:

A Programming Guide', paper prepared for the office of Women in Development, AID, Washington D.C., September.

Hoskins, Marilyn (1979b): 'Community Participation in African Fuelwood Production, Transformation and Utilisation', discussion paper prepared for Workshop on Fuelwood and other Renewable Fuels in Africa, Paris, Overseas Development Council, AID, November 29-30.

Hoskins, Marilyn (1982): 'Community Management of Natural Resources', in *Women in Natural Resources: An International Perspective*, ed. by Molly Stock *et al.*, Forestry, Wildlife and Range Sciences, University of Idaho, Moscow, USA, July.

Hoskins, Marilyn (1983): 'Rural Women, Forest Outputs and Forestry Projects', discussion draft paper No. MISC/83/3, FAO, Rome.

Howe, James W. and Staff of the Overseas Development Council (1977): *Energy for the Villages of Africa: Recommendations for African Governments and Outside Donors*, Overseas Development Council, Washington D.C., February 2.

Howes, Michael (1979): 'The Uses of Indigenous Technical Knowledge in Development', *IDS Bulletin*, UK, Vol. 10, No. 2, January.

Howes, Michael and Robert Chambers (1979): 'Indigenous Technical Knowledge: Analysis, Implications and Issues', *IDS Bulletin*, UK, Vol. 10, No. 2, January.

Hughart, David (1979): 'Prospects for Traditional and Non-conventional Energy Sources in Developing Countries', World Bank Staff Working Paper No. 346, World Bank, Washington D.C., July.

Huria, Vinod K. and K.T. Acharya (1983): 'Meeting Basic Needs Through Micro-Planning: Central Role of Essential Forestry', *Economic and Political Weekly*, Vol. 18, No. 34, August 20.

Hussain, Akmal (1980): *The Impact of Agricultural Growth on Changes in Agrarian Structure in Pakistan with Special Reference to Punjab: 1960-78*, Ph.D. dissertation, Department of Economics, University of Sussex, UK.

IADP (1966): *Intensive Agricultural Development Programme, Second Report (1960-65)*, Expert Committee on Assessment and Evaluation, Ministry of Food, Agriculture, Community Development and Cooperation, Government of India.

IADP (1969): *Modernizing Indian Agriculture: Fourth Report on the Intensive Agricultural Development Programme (1960-68) Vol. 1*, Expert Committee on Assessment and Evaluation, Ministry of Food, Agriculture, Community Development and Cooperation, Government of India.

*Indian Express* (1983): 'Tribals Hostile to Social Forestry', July 10, India.

Ishikawa, Shigeru (1975): 'The Chinese Method of Technological Development: The Case of the Agricultural Machinery and Implement Industry', *The Developing Economies*, Vol. 13, No. 4, December.

Islam, M.N. (1980): *Study of the Problems and Prospects of Biogas Technology as a Mechanism for Rural Development: Study of a Pilot Area of Bangladesh*, Final Draft Report prepared for the International Development Research Center, Ottawa, Canada, September.

Islam, Shamima (ed., 1982): *Exploring the Other Half: Field Research with Rural Women in Bangladesh*, Women for Women, Dhaka, Bangladesh.

ITES (1981): *Rural Energy Consumption in Southern India*, Institute of Techno-Economic Studies, Madras, India.

Jain, Shobhita (1984): 'Women and People's Ecological Movement: A Case Study of

Women's Role in the Chipko Movement in Uttar Pradesh', *Economic and Political Weekly*, Vol. 19, No. 41, October 13.

Jaquier, Nicolas (ed., 1976): *Appropriate Technology: Problems and Promises*, Development Center Studies, OECD, Paris.

Jha, P. Shankar (1983a): 'Farm Forestry Under Attack, Part I: Ecological Argument Unfounded', *The Times of India*, August 1.

Jha, P. Shankar (1983b): 'Outlines of a Forest Policy: Need to Enlist the Farmer', *The Times of India*, August 8.

Jodha, N.S. (1983): 'Market Forces and Erosion of Common Property Resources', paper presented at the International Workshop on Agricultural Markets in the Semi-Arid Tropics, ICRISAT, Andhra Pradesh, October.

Johnston, Bruce F. (1969): 'The Japanese Model of Agricultural Development: Its Relevance to Developing Nations', in *Agriculture and Economic Growth—Japan's Experience*, ed. by Kazushi Ohkawa, Bruce F. Johnston and Hiromitsu Kaneda, University of Princeton Press, University of Tokyo Press.

Joseph, Stephan (1980): 'Initiating and Implementing a Stove Programme', Chapter of a book under preparation, ITDG, London.

Joseph, Stephan and Yvonne Shanahan (1980): 'Designing a Test Procedure for Domestic Wood-burning Stoves', draft, ITDG Stoves Project, Intermediate Technology Development Group Ltd., London, October.

Joseph, Stephan (1983): 'A Preliminary Evaluation of the Impact of Wood-burning Stove Programmes', mimeo, FAO, Rome, June.

Joshi, Gopa (1981): 'For People-Oriented Social Forestry', *Voluntary Action*, India, January.

Kabagambe, Dennis M. (1976): 'Aspects of Resource Conservation and Utilisation: The Role of Charcoal Industry in the Kenya Economy', Working Paper No. 271, Institute of Development Studies, University of Nairobi, Kenya.

Kihl, Young Whan (1979): 'Politics and Agrarian Change in South Korea: Rural Modernisation by 'Induced Mobilisation', in *Food, Politics and Agricultural Development: Case Studies in the Public Policy of Rural Modernisation*, ed. by Raymond F. Hopkins, Donald J. Puchala and Ross B. Talbot, Westside Special Studies in Social, Political and Economic Development.

Kilvin, Joseph, E., Pradipto Roy, Fredrick C. Fliegel and Lalit K. Sen (1968): *Communication in India: Experiments in Introducing Change*, Project on the Diffusion of Innovations in Rural Societies, National Institute of Community Development Hyderabad, India.

Ki-Zerbo, Jacqueline (1981): 'Women and the Energy Crisis in the Sahel', *Unasylva*, Vol. 33, No. 11.

Knowland, Bill and Carol Ulinski (1979): *Traditional Fuels: Present Data, Past Experience and Possible Strategies*, USAID, September.

Kulkarni, Sharad (1983): 'Towards a Social Forestry Policy', *Economic and Political Weekly*, Vol. 18, no. 6, February 6.

Layard, Richard (1972): *Cost-Benefit Analysis*, Penguin Modern Economic Readings, Penguin Book Ltd., England.

Lele, Uma (1975): *The Design of Rural Development: Lessons from Africa*, The John Hopkins University Press, Baltimore and London.

Leonard, David K. (1977): *Reaching the Peasant Farmer: Organisation Theory and Practice in Kenya*, The University of Chicago Press, Chicago and London.

Leslie, A.J. (1980): 'Logging Concessions: How to Stop Losing Money', *Unasylva*, Vol. 32, No. 129.

Lewis, Oscar (1951): *Life in a Mexican Village: Tepoztlan Restudied*, University of Illinois Press, Urbana, USA.

Lipton, Michael (1968): 'The Theory of the Optimizing Peasant', *Journal of Development Studies*, Vol. IV, No. 3, April.

Lipton, Michael (1977): *Why Poor People Stay Poor: Urban Bias in World Development*, Maurice Temple Smith, London.

Little, I.M.D. and J.A. Mirrlees (1974): *Project Appraisal and Planning for Developing Countries*, Heinemann Educational Books Limited, London.

Longhurst, Richard (1980): 'Work and Nutrition in Nigeria', mimeo, Institute of Development Studies, University of Sussex, UK.

Makhijani, Arjun and Alan Poole (1975): *Energy and Agriculture in the Third World*, Ballinger Publishing Company, Cambridge, Mass., USA.

Makhijani, Arjun (1976): 'Solar Energy and Rural Development in the Third World', *Bulletin of Atomic Scientists*, June.

Makhijani, Arjun (1979): 'Economics and Sociology of Alternative Energy Sources', paper presented at the Environment and Development Regional Seminar on Alternative Patterns of Development and Life Style in Asia and the Pacific, ESCAP and UNEP, Bangkok, 14-18 August.

Malgavkar, P.D. (1981): *Rural Health Care: The Jamkhed Project*, Center for Policy Research, New Delhi, August.

Mamdani, Mahmood (1972): *The Myth of Population Control: Family, Caste and Class in an Indian Village*, Monthly Review Press, New York.

McNicoll, Geoffrey (1975): 'Community-Level Population Policy: An Exploration', *Population and Development Review*, September.

Mellor, John (1966): *The Economics of Agricultural Development*, Cornell University Press, Ithaca, New York.

Ministry of Fuel and Power (1944): *The Efficient Use of Fuel*, Her Majesty's Stationary Office, London, July.

Misra, Anupam and Satyendra Tripathi (1978): *Chipko Movement, People's Action for Development with Justice*, Gandhi Book House, New Delhi.

Misra, K.K. (1975): 'Safe Water in Rural Areas: An Experiment in Promoting Community Participation in India', *International Journal of Health Education*, Vol. 18, No. 1.

Montgomery, John D. (1965): 'The Bureaucracy as a Modernizing Elite: Can Government Routines Lead to Development', Appendix D in *Policies for Promoting Agricultural Development*, ed. by David Hapgood, Report of a Conference on Productivity and Innovation, Report No. C/65-3, Center for International Studies, MIT, Cambridge.

Moulik, T.K. (1978): 'Biogas Systems: Alternative Technology for Meeting Rural Energy Needs in India', mimeo, Second Seminar on Management Research, Indian Institute of Management, Ahmedabad, January.

Moulik, T.K. (1979): 'Strategies for Implementation of Rural Development Programmes in India', mimeo, Indian Institute of Management, Ahmedabad, India, March.

Myres, Norman (1978): 'Forests for People', *New Scientist*, 21/28 December.

Nagbrahman D. and Shreekant Sambrani (1983): 'Women's Drudgery in Firewood Collection', *Economic and Political Weekly*, January 1-8.

Narayan, Sachindra (1982): 'Social Forestry in Tribal Bihar', *Voluntary Action*, February.

Nath, Devinder (1968): *Of Logs and Men*, Committee on Case Studies, Indian Institute of Public Administration, New Delhi, India.

National Academy of Sciences (1980): *Firewood Crops—Shrub and Tree Species for Energy Production*, National Academy of Sciences, Washington D.C.

NCAER (1981): *Report on Rural Energy Consumption in Northern India*, Environment Research Committee, National Council of Applied Economic Research, New Delhi.

Ninan, Sevanti (1983): 'Protecting the Forester', *Indian Express*, 2nd November.

Nkoniki, S.R., *et al.* (1978): 'Energy Systems and Development in Tanzania', paper presented in the Tanzania National Seminar on Science and Technology for Development, Dar es Salaam, February.

Norman, Henry (1981): 'A State-of-the-Art Report on Woodstoves in the Sahel', Background Paper No. 2, Volunteers in Technical Assistance, USA, June.

Novak, K. and A. Polycarpou (1969): 'Sociological Problems and Asian Forestry', *Unasylva*, FAO, Vol. 23 (3), No. 94.

Omvedt, Gail (1984): 'Ecology and Social Movements', *Economic and Political Weekly*, Vol. 19, No. 44, November 3.

Openshaw, Keith (1971): *Present Consumption and Future Requirements of Wood in Tanzania*, Technical Report 3, FOSF/TAN 15, FAO, Rome.

Openshaw, K. and J. Morris (1976): 'The Socio-Economics of Agro-Forestry', Division of Forestry, University of Dar es Salaam, Tanzania.

Openshaw, Keith (1978): 'Woodfuel—A Time of Re-Assessment', *Natural Resources Forum*, Vol. 13, No. 1, October 6.

Paddon, A.R. and A.P. Harker (1979): 'The Production of Charcoal in a Portable Metal Kiln', Report No. G. 119, Tropical Products Institute, UK.

Pearson, G.I. (1869): 'Sub-Himalayan Forests of Kumaon and Garhwal' in 'Selections from the Record of the Government of the North-Western Provinces', Second Series, Vol. II, Allahabad, India.

PEO/AMU (1976): *The Higher Yielding Varieties Programme in India (1970-75), Part II*, Project Evaluation Organisation (Planning Commission of India) and the Australian National University, Canberra.

PIDT (1982): *Towards Participation—Nepal: Case Studies of Small Farmers Development Programme*, People's Institute for Development and Training, New Delhi.

Poulsen, Gunnar, (1978): *Man and Trees in Tropical Africa*, Publication No. 101c, International Development Research Center, Ottawa.

Powell, John W. (1978): 'Wood Waste as an Energy Source in Ghana', in *Renewable Energy Resources and Rural Applications in the Developing World*, ed. by Norman L. Brown, AAAS Selected Symposia Series No. 6, Westside Press, USA.

Prasad, N.B. (1979): *Report of the Working Group on Energy Policy*, Ministry of Energy, Department of Power, Government of India, New Delhi.

Prasad, Rajendra (1982): *Cook Stoves—a Study*, Consortium on Rural Technology, Delhi.

Raju, S.P. (1953): *Smokeless Kitchens for the Millions*, The Christian Literature Society, Madras, India.

Ravindranath, N.H., H.I. Somashekhar, R. Ramesh, Amala Reddy, K. Venkatram and Amulya K.N. Reddy (1978): *The Design of a Rural Energy Center for Pura Village, Part I, Its Present Pattern of Energy Consumption*, ASTRA, Indian Institute of Science, Bangalore, India.

Reddy, A.K.N. and K. Krishna Prasad (1977): 'Technological Alternatives and the Indian Energy Crisis', *Economic and Political Weekly*, Special No., August.

Revelle, Roger (1978): 'Requirements for Energy in the Rural Areas of Developing Countries' in *Renewable Energy Resources and Rural Applications in the Developing World*, ed. by Norman L. Brown. *op. cit.*

Richards, Paul (1978): 'Community Environmental Knowledge in Rural Development', paper presented at the workshop on 'Indigenous Technical Knowledge' at the Institute of Development Studies, University of Sussex, UK, 13-14 April.

Richardson, S.D. (1966): *Forestry in Communist China*, The John Hopkins Press, Baltimore, Maryland.

Rogers, Everett M. (1961): 'Characteristics of Agricultural Innovators and Other Adopter Categories', Research Bulletin No. 882, Ohio Agricultural Experiment Station, Wooster, Ohio, May.

Rogers, Everett M. and F. Floyd Shoemaker (1971): *Communication of Innovation: A Cross-Cultural Approach*, The Free Press, New York.

Rogers, Everett M. (1977): 'Network Analysis of the Diffusion of Innovations: Family Planning in Korean Villages', in *Communication Research—A Half-Century Appraisal*, ed. by D. Lerner and L.M. Nelson, East-West Center, Hawaii.

Rogers, Everett M. (1980): 'The Diffusion of Technical Innovations: Applications to Renewable Energy Resources in Developing Areas', paper prepared for the Panel on the Introduction and Diffusion of Renewable Energy Technologies, National Academy of Sciences, Washington, D.C., November.

Romm, Jeff (1979): 'Local Organisations for Managing Natural Resources: The Distribution of Economic Incentives and Social Shares', mimeo, Ford Foundation, Delhi.

Romm, Jeff and David Secklar (1979): Intensive Fuelwood Cultivation (IFC) or Fuel Gardening', mimeo, Ford Foundation, New Delhi, July 15.

Rosenberg, Nathan (1975): 'Factors Affecting the Payoff to Technological Innovations', mimeo, Science Policy Research Unit, University of Sussex, UK.

Roy, Burman B.K. (1979): 'Forestry in the Himalayas: For the People and by the People', in *Man and Forest: a New Dimension in the Himalaya*, ed. by Krishnamurti Gupta and Deshbandhu, Today and Tomorrow Printers and Publishers, India.

Roy, S. (1980): 'Experiences in Community Forestry, Behro Project in Bihar' in *Community Forestry Management for Rural Development*, ed. by R.N. Tewari and O.A. Mascarenhas, Papers and Proceedings of the Second Workshop, Nataraj Publishers, Dehradun.

Sandoval, Francisco Javier (1983): 'Was it the Wrong Stove?", *UNICEF News*, Issue 117.

Sansom, Robert L. (1969): 'The Motor Pump: A Case Study of Innovation and Development', *Oxford Economic Papers*, Vol. 21, No. 1, March.

Sarin, S. (1980): 'Experiences in Community Forestry: Madhya Pradesh' in *Community Forestry Management for Rural Development*, ed. by R.N. Tewari and O.A. Mascerenhas, *op. cit.*

Sarin, Madhu (1981): *Chulha Album*, mimeo, The Ford Foundation, Delhi, July.

Sarin, Madhu (1983): *Nada Chulha: Chulha Mistri's Manual*, 48 Sector 4, Chandigarh,

India, September.

Sarin, Madhu and Uno Winblad (1983): 'Cook Stoves in India—A Travel Report', mimeo. Swedish International Development Agency, Delhi, May.

Schofield, Sue (1979): *Development and the Problems of Village Nutrition*, Croom Helm, London.

Schultz, Theodore (1964): *Transforming Traditional Agriculture*, Yale University Press, New Haven, USA.

Sen, Amartya, K. (1970): *Collective Choice and Social Welfare*, Holden-Day, Inc., California, USA.

Sen, Amartya K. (1972): 'Feasibility Constraints: Foreign Exchange Shadow Wages', in *Cost-Benefit Analysis*, ed. by Richard Layard, Penguin Books, UK.

Sen, Amartya (1973): *On Economic Inequality*, Oxford University Press, India.

Sen, Lalit K. (1969): *Opinion Leadership in India: A Study of Interpersonal Communication in Eight Villages*, National Institute of Community Development, Hyderabad, India.

Sethi, Harsh (1980): 'Alternative Development Strategies—A Look at Some Micro Experiments', in *Readings in Poverty, Politics, and Development*, ed. by Kamla Bhasin and Vimla R., Freedom from Hunger Campaign/Action for Development, FAO, Rome.

Shaller, Dale V. (1979): 'Socio-Cultural Assessment of the Lorena Stove and its Diffusion in Highland Guatemala', mimeo, VITA, Maryland, USA.

Shephard, K.R. (1979): 'Energy for the Forests: An Exercise in Community Forestry for Developing Countries', in *Forestry in National Development: Production Systems, Conservation, Foreign Trade and Aid*, ed. by K.R. Shephard and H.Y. Richter, Development Studies Center Monograph No. 17, Australian National University.

Shiva, Vandana and Jayanto Bandhyopadhyay (1984): *Ecological Audit of Eucalyptus Cultivation in Rainfed Regions*, A Report for the United Nations University, Research Foundation for Science, Technology and Natural Resource Policy. Dehradun, India.

Singer, H. (1961): 'Improvement of Fuelwood Cooking Stoves and Economy of Fuelwood Consumption', Report No. TA 1315, FAO, Rome.

Siwatibau, Suliana (1978): *A Survey of Domestic Energy Use and Potential in Fiji*, Center for Applied Studies in Development, University of the South Pacific, Fiji.

Skar, Sarah Lund (1982): 'Fuel Availability, Nutrition and Women's Work in Highland Peru', World Employment Programme Research Paper No. WEP 10/WP 23, International Labour Office, Geneva.

Smil, Vaclav (1979): 'Energy Flows in the Developing World', *American Scientist*, Vol. 67, No. 5, Sept-Oct.

Smil, Vaclav and William E. Knowland (1980): 'Energy in the Developing World' in *Energy in the Developing World: The Real Energy Crisis*, ed. by the same authors, Oxford University Press, UK.

Smith, K.R., A.L. Agarwal and R.M. Dave (1983): 'Air Pollution and Rural Fuels: Implications for Policy and Research', Resource Systems Institute, East-West Center, Honolulu, Hawaii.

Smith, William A. (c. 1973): 'Concientizacao and Simulation Games'. Technical Note No. 2, The Ecuador Non-Formal Education Project, Ministry of Education, Ecuador, and the Center for International Education at the University of

Massachusetts, USA.

Smythies, E.A. (1925): *India's Forest Wealth*, London.

Spears, John S. (1978): 'Wood as an Energy Source: The Situation in the Developing World', paper for the 103rd Annual Meeting of the American Forestry Association, USA, October 8.

Squire, L. and M.G. Van der Taak (1975): *Economic Analysis of Projects*, John Hopkins Press, USA.

Staudt, Kathleen, A. (1976): 'Women Farmers and Inequalities in Agricultural Services'. *Rural Africana*, No. 29, Winter, African Studies Center, Michigan.

Stewart, Frances (1975): 'A Note on Social Cost-Benefit Analysis and Class Conflict in LDCs', *World Development*, Vol. 3, No. 1, January.

Stock Molly, Jo Ellen Force and Dixie Ehrenreich (eds., 1982): *Women in Natural Resources: An International Perspective*, Forest, Wildlife and Range Experiment Station, University of Idaho, Moscow, USA.

Stone, Linda (1982): 'Women in Natural Resources: Perspectives from Nepal', in *Women in Natural Resources: An International Perspective, op. cit.*

Subramaniam, T.S. (1983): 'Tamil Nadu Loses Crores Over Forestry Projects', *Indian Express*, July 2.

Swaminathan, M.S. (1980): 'Indian Forests at the Cross Roads', in *Community Forestry and People's Participation: Seminar Report*, Ranchi Consortium for Community Forestry, November 20-22.

Swaminathan, Mridula (1984): 'Eight Hours a Day for Fuel Collection'. *Manushi* (India), March-April.

Swaminathan, Srilata (1982a): 'Environment: Tree versus Man' in *India International Center Quarterly*, Vol. 9, Nos. 3 and 4.

Swaminathan, Srilata (1982b): 'The Green Revolution Bound up in Red Tape', Report No. 82, Center for Science and Environment, New Delhi.

*Sylva Africana* (1980): 'Charcoal Production—Focus of Research', No. 7, February.

Tanzer, Michael (1974): *The Energy Crisis: World Struggle for Power*, Monthly Review Press, New York.

Torres, Augusto, Stanley Lichtenstein, Paul Spector (1968): *Social and Behavioural Impacts of a Technological Change in Colombian Villages*, Report, American Institute for Research, International Research Institute, Washington D.C.

Tuschak, T.S. (1979): 'Energy Policy in Developing Countries—Problems and Challenges', paper prepared for the Energy Forum on Third World Energy Strategies and the Role of Industrialised Countries, London, June 20-22.

Uhart, E. (1975): 'Charcoal Development in Kenya', UN Economic Commission for Africa, ECA/FAO Forest Industries Advisory Group for Africa, Ind-91/MR-29, July.

Uhart, E. (1976): 'The Wood Charcoal Industry in Africa', memorandum, African Forestry Commission, Fourth Session, Bangui, Central African Republic, 22-27, March.

UNIDO (1972): *Guidelines for Project Evaluation*, United Nations, New York.

United Nations (1984): *Energy Statistics Yearbook 1982*, Department of International Economic and Social Affairs, Statistical Office, New York.

USAID (1970): 'A Survey of India's Export Potential of Wood and Wood Products', Vol. I, Delhi.

UTTAN (1983): 'A Question—Why is Social Forestry Not 'Social'?', paper presented at the Ford Foundation workshop on Social Forestry and Voluntary Agencies, Badkhal Lake, Haryana, 13-15 April.

Van Buren, Ariane (ed., 1979): *A Chinese Biogas Manual*, Intermediate Technology Publications Ltd., UK, July.

Van Buren, Ariane (1980): *Biogas Training in China: A First Exchange with Developing Countries*, Report on the First International Biogas Training Seminar in China, United Nations Environmental Programme, Nairobi.

Verma, N.L. (1983): 'Rajasthan's Social Forestry Project—A Review', *Economic Times*, India, August.

Walton J.D. Jr., A.H. Roy and S.H. Bomar Jr. (1978): *A State of the Art Survey of Solar Powered Irrigation Pumps, Solar Cookers and Wood Burning Stoves for Use in Sub-Sahara Africa*, mimeo, Engineering Experiment Station, Georgia Institute of Technology, Atlanta, Georgia, USA, January.

Weisbrod, B.A. (1968): 'Deriving an Implicit Set of Government Weights for Income Classes', in *Cost-Benefit Analysis*, ed. by Layard, *op. cit.*

Wharton, Clifton Jr. (1971): 'Risk, Uncertainty, and the Subsistence Farmer' in *Economic Development and Social Change*, ed. by George Dalton, Natural History Press Garden City, New York.

White, Benjamin (1976): 'Population, Involution and Employment in Rural Java'. in *Agricultural Development in Indonesia*, ed. by Gary E. Hansen, Cornell University Press, USA.

White, Benjamin (1984): 'Measuring Time Allocation, Decision-Making and Agrarian Changes Affecting Rural Women: Examples from Recent Research in Indonesia', in *Research on Rural Women: Feminist Methodological Questions, IDS Bulletin*, Vol. 15, No. 1, Institute of Development Studies, Sussex, UK.

Williams, Paula J. (1982): 'Women and Forest Resources: A Theoretical Perspective' in *Women in Natural Resources: An International Perspective, op. cit.*

Wood, T.S. (1982): 'Improved Woodstoves in the Sahel—A Critical Assessment', proceedings of a workshop on Energy, Forestry and Environment, USAID, Africa Bureau.

World Bank (1983a): *Gujarat Community Forest Project—Mid-Term Review Mission Report*, World Bank, India.

World Bank (1983b): *Uttar Pradesh Social Forestry Project—Mid-Term Review Mission Report*, World Bank, India.

World Bank, (1984): *World Development Report 1984*, Washington D.C.

Wyon, John B. and John E. Gordon (1971): *The Khanna Study: Population Problems in Rural Punjab*, Harvard University Press, Massachusetts, USA.

Zumer-Linder, Majda (1976): 'Firewood Crisis and Village Forestry', mimeo, International Youth Federation for Environmental Studies and Conservation, Denmark, March.

# AUTHOR INDEX

# SUBJECT INDEX

# INSTITUTE OF ECONOMIC GROWTH

# Studies in Economic Development and Planning

1. *The Role of Small Enterprises in Indian Economic Development* by P.N. Dhar and H.F. Lydall.
2. *Indian Tax Structure and Economic Development* by G.S. Sahota
3. *Agricultural Labour in India* by V.K.R.V. Rao.
4. *Foreign Aid and India's Economic Development* by V.K.R.V. Rao and Dharam Narain.
5. *Agricultural Production Functions, Costs and Returns in India* by C.H. Hanumantha Rao.
6. *Resource Allocation in the Cotton Textile Industry* by Dharma Kumar, S.P. Nag and L.S. Venkataramanan.
7. *India's Industrialisation and Mineral Exports* by V.P. Chopra.
8. *Taxation of Agricultural Lands in Andhra Pradesh* by C.H. Hanumantha Rao.
9. *Some Aspects of Cooperative Farming in India* by S.K. Goyal
10. *Demand for Energy in North-West India* by P.N. Dhar and D.U. Sastry.
11. *Some Aspects of the Structure of Indian Agricultural Economy 1947-48 to 1961-62* by P.V. John.
12. *Wages and Productivity in Selected Indian Industries* by J.N. Sinha and P.K. Sawhney.
13. *Inventories in Indian Manufacturing* by K. Krishnamurty and D.U. Sastry.
14. *Buffer Stocks and Storage of Major Foodgrains in India* by A.M. Khusro.
15. *Bangladesh Economy: Problems and Prospects* ed. by V.K.R.V. Rao.
16. *The Economics of Land Reforms and Farm Size in India* by A.M. Khusro.
17. *Technological Change and Distribution of Gains in Indian Agriculture* by C.H. Hanumantha Rao.
18. *Investment and Financing in the Cooperative Sector in India* by K. Krishnamurty and D.U. Sastry.
19. *Land Reforms in India: Trends and Perspectives* by P.C. Joshi.
20. *Cost Benefit Analysis: A Case Study of the Ratnagiri Fisheries* by S.N. Mishra and John Beyer.
21. *Livestock Planning in India* by S.N. Mishra.
22. *Rice Marketing System and Compulsory Levies in Andhra Pradesh* by K. Subbarao.
23. *Farm Size, Resource-Use Efficiency and Income Distribution* by G.R. Saini